ONE WEEK LOAN

1 7 OCT 2007

0 3 MAY 2011

Embedded Microprocessor Systems
Real World Design

Second Edition

Stuart R. Ball

Newnes
An imprint of Butterworth-Heinemann

Boston Oxford Auckland Johannesburg Melbourne New Delhi

Newnes is an imprint of Butterworth–Heinemann.

Library of Congress Cataloging-in-Publication Data
Ball, Stuart R., 1956–
   Embedded microprocessor systems : real world design / Stuart R. Ball.
      p.   cm.
   Includes index.
   ISBN 0-7506-7234-X (pbk. : alk. paper)
   1. Embedded computer systems—Design and construction.
2. Microprocessors.   I. Title.
TK7895.E42 B35 2000
004.16—dc21   99-086910

British Library Cataloguing-in-Publication Data
A catalogue record for this book is available from the British Library.

The publisher offers special discounts on bulk orders of this book.
For information, please contact:
Manager of Special Sales
Butterworth-Heinemann
225 Wildwood Avenue
Woburn, MA 01801-2041
Tel: 781-904-2500
Fax: 781-904-2620

For information on all Butterworth–Heinemann publications available, contact our World Wide Web home page at: http://www.newnespress.com

10 9 8 7 6 5 4 3 2

Printed in the United States of America

Contents

Introduction

Imagine this scene: You get into your car and turn the key on. You take a 3.5″ floppy disk from the glove compartment, insert it into a slot in the dashboard, and drum your fingers on the steering wheel until the operating system prompt appears on the dashboard liquid crystal display (LCD). Using the cursor keys on the center console, you select the program for the electronic ignition, then turn the key and start the engine. On the way to work you want to listen to some music, so you insert the program compact disc (CD) into the player, wait for the green light to flash indicating that the digital signal processor (DSP) in the player is ready, then put in your music CD.

You get to work and go to the cafeteria for a pastry. Someone has borrowed the mouse from the microwave but has not unplugged the microwave itself, so the operating system is still up. You can heat your breakfast before starting work.

What is the point of this inconvenient scenario? This is how the world would work if we used microprocessor technology without having *embedded microprocessors*. Every microprocessor-based appliance would need a disk drive, some kind of input device, and some kind of display.

Embedded microprocessors are all around us. Since the original Intel 8080 was pioneered in the 1970s, engineers have been embedding microprocessors in their designs. They even are embedded in general-purpose computers; if you own a variation of the IBM PC/AT, there is an embedded microprocessor in the keyboard. Virtually all printers have at least one microprocessor in them, and no car on the market is without at least one under the hood. Embedded microprocessors may control the automatic processing equipment that cans your soup or the controls of your microwave oven.

Basically, we can define an embedded microprocessor as having the following characteristics:

Dedicated to controlling a specific real-time device or function.

Self-starting, not requiring human intervention to begin. The user cannot tell if the system is controlled by a microprocessor or by dedicated hardware.

Self-contained, with the operating program in some kind of nonvolatile memory.

Of course, there are exceptions to this general description, which we will get to eventually, but this definition will serve us for now.

An embedded microprocessor system usually contains the following components:

A microprocessor

RAM (random access memory)

Nonvolatile storage: erasable programmable read-only memory (EPROM), read-only memory (ROM), flash memory, battery-backed RAM, and so forth

I/O (some means to monitor or control the real world)

Of course, if you have seen textbooks describing general computer systems, this description fits those as well. The difference is in the details. A general-purpose computer, such as the one this book was written on, may have many megabytes of RAM, whereas an embedded system may have less than 256 bytes (that is bytes, not megabytes) of RAM. Your PC at home or at the office may have a 10 GB IDE hard drive with DOS, Windows, and several other applications. An embedded system usually contains its entire program in a few thousand bytes of EPROM. The most important difference between the two is the application. Your home personal computer (PC) runs a word processor, then you switch over to the money management program to balance the checkbook, then to the spreadsheet to work on the family budget, then back to the word processor. The embedded system does just a limited number of tasks, such as making sure your toast does not burn or timing the cook cycle in your microwave.

Why would anyone want to use a microprocessor? The main reasons are:

- **Cost.** The cost of developing firmware for an embedded system can be very high, but it is a *nonrecurring expense* (NRE), only spent once to develop the product. The actual cost of the finished product can be very low. On the other hand, the product cost of a system such as a microwave oven controller, if implemented in discrete hardware, can be very high by comparison.
- **Flexibility.** Say a typical microwave oven manufacturer gets a contract from a very large discount store for microwave ovens, but the contract specifies certain changes in the way the user controls the device. In a hardware-based system, the control electronics would need to be redesigned.

In a microprocessor-based system, the only change may be a few lines of code.

- **Programmability.** You may want to program a robotic arm to paint car doors one day and trunk lids the next. The proper embedded microcontroller permits you to have the same hardware perform different tasks. Of course, this also could be implemented in discrete hardware but at much higher cost.

This book will take you step by step through the procedures involved in designing an embedded control system. Many of the tricks I have learned in my 20 years in the field will be passed on, as well as some pitfalls to avoid. Along the way, we will use as an example a simple embedded control system, a swimming pool pump timer, to illustrate the concepts.

The book is aimed primarily at students, new graduates who will be moving into the embedded processor field, and engineers working in another field who want to switch to embedded microprocessors. It assumes that the reader has a basic knowledge of software concepts, binary and hexadecimal number systems, and a basic understanding of digital logic. A review of this material is included in the appendixes at the end.

System Design 1

It has been said that if you do not know where you are going, you will not know when you get there. Success experts tell us that the first step in achieving anything is to establish a goal—to be debt free in one year or to pay off the car in six months.

Like most things in life, the process of designing an embedded microprocessor system begins with a goal—the definition of the product. The product definition describes what the product is to be and do.

The product definition is the first element in a process that is key to any successful electronics system design: documentation. The documentation describes what you are going to build and how you are going to build it. It tells marketing people what product they will have to sell, and it tells the engineering team how to implement that product. Since this book is about embedded systems, it will focus on documenting embedded systems. The development documents that I have found useful in developing embedded systems are as follows:

Product requirements definition

Functional requirements

Engineering specifications

Hardware specifications

Firmware specifications

Test specifications

Briefly, the product requirements describe what the product is. The functional requirements describe what the product must do. The engineering specifications describe how the design will be implemented and how the requirements will be met. The hardware specifications describe how specific hardware is designed, and similarly, the firmware specifications describe how the firmware for specific processors will be designed. Finally, the test specifications describe what must be tested and how to verify that the system operates

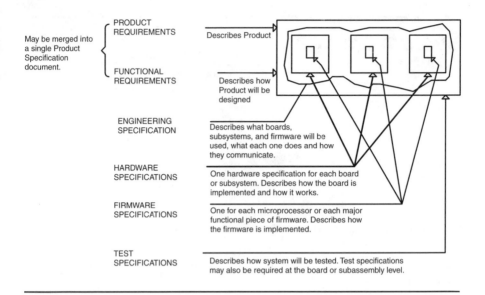

Figure 1.1
Design Documentation.

correctly. Figure 1.1 shows how each of the documents relates to the overall design. The embedded design process generally follows these steps:

Product requirements definition

Functional requirements definition

Processor selection

Hardware/software specifications

System evaluation

Hardware design

Firmware design

Integration

Verification (test)

These steps are not necessarily serial. For example, if there are separate hardware and software teams, the hardware and firmware design can proceed in parallel. The process is not always linear—system evaluation may reveal a problem with the selected processor, which means that step has to be repeated. Last, the process is not always this well divided. The requirements definition and functionality description, for example, may be merged into a product specification or other customer-required documents.

Many companies require such product specifications early in the design process. I will not dwell on that here, as the requirements for this type of document are specific to the company or the customer for whom the product is intended. Commercial customers, to pick one example, have considerably different requirements than the Department of Defense.

The design and documentation process begins with the next level of documentation below the product specification: the requirements definition.

Requirements Definition

The requirements definition (which, again, may actually be part of the product specifications), describes what the product is to do. In a very large company, the requirements may be defined by the marketing department or by a major customer. In a smaller company, the hardware and software engineers may sketch it out on napkins across the lunch table. For a small, one-engineer project, the requirements may be the result of a momentary inspiration.

The requirements definition can take the form of a book, defining every interaction, interface, and error condition in the system; or it can take the form of a single-page list of what the finished product must do. In either case, the requirements definition must describe:

What the system is to do

What the real world I/O consists of

What the operator interface is (if any)

In a small embedded control system, defining the requirements is crucial, as it prevents problems later when you find out that there is insufficient RAM or that the microprocessor you have chosen is too slow for the job. A simple example of this is the system definition for a swimming pool pump timer below. (Appendix A contains the complete requirements definition and specifications.)

System Description: A swimming pool timer that cycles the alternating current (AC) pump motor on a swimming pool.

Power input: 9–12V DC from a wall-mount transformer.

Pump is a ½ hp, single-phase, AC motor, controlled by mechanical relay.

Provision is to be made for a switch closure input that inhibits pump operation if the water level is low.

User can set the length of time the pump is on and the length of time it is off. An override is available to permit turning off the pump when it is on for maintenance and turning on the pump when it is off so that chemicals can be added.

On/off/override time is to be adjustable in 30-minute increments from $\frac{1}{2}$ hour to 23 hours.

A display will indicate the on/off condition of the pump, the time remaining, and if the pump is in override mode. The display also will indicate the condition of the water-low monitor.

Minimum switches and knobs.

In addition to a list of requirements and functions like this, a system that is intended to be a commercial product might also include requirements for EMI/EMC (electromagnetic interference/electromagnetic compatibility) certification, safety agency approval (UL/IEC), and some kind of environmental (temperature, humidity, salt spray, etc.) specifications.

Although this will be discussed further in Chapter 6, one problem with specifying requirements is verifying them. It is easy to tell if the product meets the EMI/EMC requirements—you will run tests to prove it. How do you prove you've met the requirement for "Minimum switches and knobs"? Keep in mind the problem of verification when specifying requirements.

A complex system may have another level of documentation, which I usually title the *Engineering Specification*. This document describes the approach that will be used to implement the design, including which boards will be included and how the functions are partitioned onto those boards. I will return to this information later, in Chapter 7. For now, assume that we have a simple product, which makes this intermediate document unnecessary.

After the requirements are defined, the next step is to determine if a microprocessor is the best choice. For the pool timer, it is fairly obvious that a microprocessor is the easiest way to do the job. Some other systems are not so obvious. The following questions can help determine if a microprocessor is justified:

- At what speed must the inputs and outputs be processed or updated? Although the clock rates are ever increasing, there is a practical upper limit to the speed at which a microprocessor can read an input or update an output and still do any real work. At the time of this writing, an update rate of approximately 50 kHz is a practical upper limit for a simple microprocessor system with few processing demands and running on a fast processor or digital signal processor (DSP). If the system has to do significant processing, buffer manipulation, or other computing, the potential update rate will go down.

- Is there a single integrated circuit (IC) or a programmable logic device (PLD) that will do the job? If so, a microprocessor probably is not justified.
- Does the system have a lot of user I/O, such as switches or displays? If so, a microprocessor usually makes the job much easier.
- What are the interfaces to other external systems? If your system has to talk to something else using Synchronous Data Link Control (SDLC) or some other complex communication protocol, a microprocessor may be the only practical choice.
- How complex is the computational burden on the system? Modern electronic ignition systems, for example, have so many inputs (air sensors, engine rpm, etc.) with complex relationships that few choices other than a microprocessor are suitable.
- Will the design need to be changed once it is finished or will the requirements be changing as the design progresses? Is there a need for customization of the product or for special versions? Any of these requirements make a microprocessor attractive, due to the flexibility of implementing functionality in firmware.

Fortunately, the job of the system designer is becoming easier. Microprocessor costs are coming down as speed and performance goes up. Even simple microprocessors are capable of handling tasks that were limited to dedicated hardware just a few years ago. When you include very fast processors (such as low cost DSPs), the range of potential applications that can be performed with a microprocessor is wider than ever.

Processor Selection

Suppose you decide to use a microprocessor for your new widget. What steps are taken for selecting the processor to be used? Fortunately, for all but a very few applications, more than one right solution is possible because several microprocessors could meet the requirements. As with most real world engineering decisions, the selection consists of a series of trade-offs between cost and functionality. The specific selection process will depend on the complexity of the finished product, but the following items must be taken into consideration:

Number of I/O pins required

Interfaces required

Memory requirements

Number of interrupts required

Real-time considerations

Development environment

Processing speed required

ROMability

Memory architecture

Number of I/O Pins

In a minimum-cost system, component count is a major factor in the final product cost. These systems generally use a single-chip microprocessor with internal ROM and RAM. There is a convention to identify these parts as *microcontrollers*, to separate them from the more general-purpose embedded processors. Since the microcontroller does not need to generate signals to external memory, the device pins are available for I/O. These pins are grouped as *ports*, and each pin may be an input or an output. In our example system, one pin might turn on the pool pump relay. Another pin might allow the processor to monitor the water level sensor.

Most microprocessor manufacturers make a controller with internal memory and external pins for controlling I/O devices. Some make nothing else. While it is impossible to list all the variations and subtleties of these devices here, a brief list of typical devices follows:

Manufacturer	Processor	I/O Pins
Intel and others	8031/8051 family	32
Microchip	PIC17C42	33
Motorola	68HC05 family	Varies
Zilog	Z8 (Z86E40)	32
Signetics/Philips	83C751	19
Atmel	AT90S4414	32

This list cannot describe all the trade-offs among the various parts. Some of these parts, for example, include a bidirectional serial port, but you have to give up two port pins to use it. Some have internal timers that use a port pin for certain functions. Some have high-current and open-drain outputs that are ideal for driving relay or solenoid coils with no additional driver hardware. The specific IC that is ideal for your system depends on your system.

When counting I/O pins, make sure that you take into account the use of internal functions, such as the serial ports and timers, that restrict use of certain pins. Although it will be covered in more detail in Chapter 2, remember that some of these parts support external RAM or ROM, but using that capability takes anywhere from 8 to 19 I/O pins to access the external memory.

Interfaces Required

The entire point of an embedded processor is to interact with some piece of real world hardware. Not only must the hardware be in place to handle the interface, but the processor must be fast enough to perform whatever processing must be done on the data. In a single-chip system, processor selection may be highly dependent on the interface requirements. For example, the Microchip PIC17C42 has two pulse-width modulation (PWM) outputs that simplify design of such things as antilock braking systems and motor servos. One caveat: Study the data sheets carefully. Many processors have limitations that are not immediately obvious. You might find, say, that the serial port is specified as being able to operate at a certain maximum baud rate, but careful examination of the data sheet may reveal that not all modes of operation are available at the maximum rate.

Determining if a particular processor can keep up with the interface requirements is not always easy. Unfortunately, there is no magic formula to determine this. I have frequently resorted to writing part of the code for an interface just to be sure that the processor has enough capacity.

Memory Requirements

Determining the memory requirements is a key part of embedded system design. If you overestimate the memory required, you may select an unnecessarily expensive solution. If you underestimate, you risk project delays while the system is redesigned. Since memory comes only in sizes that are addressable with digital bits, such as $8\,K \times 8$, $32\,K \times 8$, and so on, you need not estimate memory requirements down to the last byte. You do need to ensure that you have enough memory.

RAM   RAM is fairly straightforward to estimate. The number of variables plus the sum of all internal buffers, FIFOs (first in, first out), and stacks is the amount of RAM required. Many single-chip microcontroller ICs are limited to less than 1024 bytes (1 K, or kilobyte) of internal memory. If the memory goes beyond what is internally available, then external RAM must be added. However, this requires the use of I/O pins to address the added memory and often defeats the purpose of using a single-chip controller.

One caution is important: Some microcontrollers have restrictions on RAM usage, such as the need to use part of the internal RAM for banks of internal registers. For a couple of examples, look at the 8031, which has 128 bytes of internal RAM. The 8031 has four register banks that use 32 bytes of that, leaving 96 usable bytes of RAM. If your application needs only one or two register banks, the rest is available for general use. The 8052 processor has 256 bytes of general-purpose RAM, but the upper 128 bytes are accessible only by using indirect addressing. The Atmel AVR90S4414 has 32 general-purpose registers, but only 16 can be used with the immediate data instructions.

The size of RAM required also will vary with the development language used. Some inefficient compilers use enormous amounts of RAM.

ROM   The amount of ROM required for a system is the sum of the program code and any ROM-based tables required. Examples of ROM tables are step motor ramp tables, data translation lookup tables, and indirect branch tables. The tables usually are straightforward to estimate. The difficult part is estimating the code size. Estimates of code size become more accurate with increasing experience, usually gained by being wrong. However, it is important to remember that being precise is not as important as knowing the upper limit on code size. One rule of thumb is if the ROM is more than 80 percent full, it is too full. Unless you can guarantee that the system requirements will never change, leave some margin. In many cases, it is worthwhile to write portions of the code, just to see how big they will get. In microcontroller-based systems with internal ROM, you are limited to whatever program memory the part contains.

Like RAM usage, code size depends somewhat on the development (programming) language selected. A program written in assembler takes less space than one written in Pascal, for example. Again, this depends on the language and even on the specific brand of software.

It is not a good idea to let the language drive the design, at least in low-cost systems. The languages easiest to use, debug, and maintain are often those that require the most memory and the most processing speed. Choosing the wrong language can turn a simple, inexpensive, single-chip design into something that requires an embedded 64-bit powerhouse with megabytes of RAM. However, sometimes company policy or a customer contract specifies the use of a high-level language. In these cases, you just have to live with the increased cost and complexity that this implies.

A real life example will illustrate the potential problems you can run into here. An embedded system was to be controlled by an x86-family processor. We had settled on an off-the-shelf CPU board, based on a 386SX. Then one of the software people noticed that the 386SX has no floating-point coprocessor (FPU). The software engineers were from the PC world, where everything ran in Windows 95/98, on a 400 MHz Pentium. They couldn't conceive of not having hardware for floating-point calculations. The only way to get a hardware floating point was to go up to a 486DX or Pentium processor, which doubled the cost of the CPU board. This was an *embedded* application, with no keyboard, display, or hard drive attached. The CPU was reading sensors, controlling motors, and communicating with a PC host. There was no reason to believe that floating-point calculations would ever be needed. But, because C makes it easy to define floating-point variables, they were expected to be available in hardware. In fact, the code wasn't designed or written yet, so we didn't *know* whether any floating-point calculations would be required.

This same design had some embedded microcontrollers for very low-level functions. What if a software engineer had decided that those needed hardware floating point and a deep stack for recursion? We'd have turned a requirement for a cheap 8-bit microcontroller into Pentium-class overkill.

Number of Interrupts Required

This will be treated in more detail in Chapter 4. However, a few comments are worth mentioning here.

Many designers overuse interrupts. An interrupt does just that—it interrupts program execution. Interrupts are best used for those things that cannot wait for the processor to get to them. In some cases, an interrupt can be used just to reduce the hardware complexity (and the associated costs), but almost always it is at the expense of increased debug time and higher potential for hard-to-find intermittent errors. In those cases where interrupts are required, it is important to know how many really are needed.

Interrupts are used to notify the processor of special events such as a timer that timed out or a piece of hardware that needs attention. Counting the events that need interrupts is straightforward, but be sure to take into account internal interrupt sources as well. Some tricks can be played to reduce the number of interrupt *signals* required when there are more interrupt *sources* than the processor has interrupt inputs. Again, these will be covered in Chapter 4.

Real-Time Considerations

This subject covers a lot of territory and is closely connected to the issue of processing speed.

Real-time events are what embedded microprocessors generally are intended to handle. However, some specific events deserve special consideration. For example, you might have a subsystem that controls a motor using

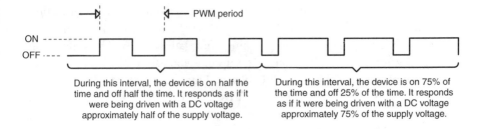

During this interval, the device is on half the time and off half the time. It responds as if it were being driven with a DC voltage approximately half of the supply voltage.

During this interval, the device is on 75% of the time and off 25% of the time. It responds as if it were being driven with a DC voltage approximately 75% of the supply voltage.

Figure 1.2
PWM Operation.

pulse-width modulation. In this scheme, the motor current is controlled by switching the current at a very high rate and using the duty cycle to control the motor speed. The motor, being a relatively slow mechanical device, responds to the time-average of the current (see Figure 1.2). Lower duty cycles result in lower average current and slower rotation. (This is a very high-level description; entire books have been written about PWM and motor control. Read one of those for all the details.)

In our hypothetical motor-control system, say that the microprocessor cannot keep up with the motor on a real-time basis. That is, the *chopping* rate, the rate at which the motor current is switched on and off, is faster than the microprocessor can handle. But the other required tasks, such as communicating with whatever is controlling the motor-processor subsystem, are no problem for our processor. It seems that we must go to a much faster, more expensive processor to keep up with the motor, thus raising the cost of the system.

There is another solution, however. Many microprocessors have PWM outputs or timers that can be configured to operate as PWM outputs. Typical examples are the Microchip PIC 16C/17C family, the Atmel AT90S family, and the Intel 80C196 series. Using the internal PWM controller relieves the microprocessor of the burden of generating every motor current change. Instead, the processor just sends *changes* in the duty cycle (or frequency) to the PWM controller.

This is just one example of how picking the right processor can solve a real-time problem. Other examples include selecting a processor with a built-in, high-speed serial port for interprocessor communications; selecting a processor with an on-chip direct memory access (DMA) controller (more about that in a later chapter); or selecting a processor with special memory manipulation registers that will speed things up.

Development Environment

The development environment often is a key consideration. New development tools require a learning curve, and with a tight development schedule there often is no time to research, acquire, and become proficient with a new set of tools. If the company has several tens of thousands of dollars (a not unrealistic figure) invested in emulators for a specific processor and if all the software engineers are comfortable with those tools, someone usually objects to changing processors just so an enthusiastic engineer can tinker with the latest chip. That is not much fun for the frustrated engineer, but it is an economic fact of life. This is why some companies (or subsidiaries within very large companies) expend a great deal of effort to pick a processor family they can live with for a long time.

Even if a design starts with a blank slate, however, the development tools can be a major consideration. For example, selecting a widely used processor, such as the 8031, allows you to select from a wide array of tools from a number of vendors. The capability of these tools (such as emulators) can be matched to whatever budget you have. On the other hand, the tools for some specialized processors are available only from the manufacturer and the cost can be prohibitive.

Tools can be a major factor. If the processor choice gets down to just two, researching the cost of tools may make the decision obvious. In any event, be sure you know the cost of these tools, especially emulators from the IC manufacturer, before you make the final selection.

If you're planning to use an RTOS (real-time operating system), the choice of which one to use also may drive your processor selection. RTOSs come in various flavors, with some charging a one-time fee and others charging a license fee or royalty for every unit you build. Some have a flat royalty, some charge a little for every module you include. I worked on a system once where one engineer wanted to embed an RTOS in four of the processors. We'd have spent around $800 per system just in RTOS license fees. Make sure you choose a processor for which a suitable RTOS is available and that the RTOS costs are compatible with your product cost.

Processing Speed Required

This is another area that is easier to get right after you have done it for a while, but a few guidelines can help:

- Add up the interrupt latencies. The processor has to be fast enough that a worst-case stack-up of interrupts (it *will* happen) can be handled without anything bad occurring. We'll return to this in the chapter on interrupts.

- The length of the polling loop (more about that in a later chapter) must be short enough to never miss a byte of serial data or a byte from any other interface. In an interrupt-driven system, the same considerations apply to the length of any polling loop plus the worst-case interrupt latencies.
- Note that, in some cases, going to higher speeds gains nothing if wait states have to be inserted to meet the memory access time requirements. We'll look at wait states in Chapter 2.

Common pitfalls about processor speed are as follows:

Confusing clock rate with processor speed. A standard 8031, for example, can accommodate an input clock of 12 MHz. So it's a 12 MHz processor, right? Wrong. The clock circuitry divides the clock by 12 because the internal logic needs several phases, or clock edges, per instruction. This yields a processor rate of 1 MHz. Many processors such as the 80186/80188 divide the external clock by 2. The PIC-family processors divide the clock by 4, but the Atmel AT90S series parts do not. So, at least in raw execution speed, an 8 MHz AT90S part (8 MHz clock, 8 MHz execution rate) is faster than a 20 MHz PIC part (20 MHz clock, 5 MHz execution rate). None of these characteristics is bad unless you do not know them or do not take them into account.

Not evaluating the instruction set. The Atmel AT90S and Microchip PIC 16C/17C series parts have a fairly high execution speed. However, the reduced instruction set computer (RISC) architecture can be a real trap. For example, these parts lack sophisticated indirect (lookup table) branch capability. An indirect branch function can be constructed, but that takes some instructions. Similarly, the parts only have one branch instruction (GOTO). Conditional branches require two or more instructions to construct. Consequently, the potential execution rate is reduced by the extra code involved in manipulating the hardware. A RISC microcontroller may execute instructions very fast, but it may not be as fast as a CISC (complex instruction set computer) with an instruction set that can perform complex operations. For example, multiplying two 16-bit numbers may take one instruction and only a few clock cycles on a CISC processor or a single cycle on a DSP with multiplier hardware. On a RISC processor that has no multiply instruction or multiply hardware, this operation will have to be implemented in some kind of loop that uses several instructions and a large number of clock cycles.

Not evaluating the architecture. The ADSP-2100 family parts from Analog Devices are DSPs that lend themselves well to embedded applications.

These parts are optimized for signal processing, which means that they have some powerful data manipulation capabilities such as hardware multiply and barrel shifters. However, they also have some limitations. Some operations require an extra instruction to move a value from RAM to a register before it can be used, where other, slower processors allow the value in RAM to be manipulated or tested directly.

These are typical and by no means unique. Every processor has its quirks. These things are not dark secrets. You just have to understand the data sheets on the part before you use it. Take the data book or CD-ROM home. Read it. Study the timing diagrams, especially the worst-case numbers. Understand how everything in your system will connect to and be controlled by this processor. If you do not understand something, you are not ready to start the design.

ROMability

This consideration applies only to those devices that execute their programs from internal ROM. These devices usually are chosen for an application where cost, rather than being no object, is a key factor. If the finished design is going to be a very high-volume (thousands per year) product, it may be worthwhile to select a processor that has a ROM version.

Most engineering projects use EPROM or flash memory for their development phases. These erasable and reprogrammable memories allow a part to be reused instead of thrown away. When the part goes to production, the EPROM parts can be replaced with one-time programmable (OTP) devices. These usually are just EPROM-based parts in a plastic package and with no erasure window. Since the expensive ceramic package and quartz window are not required, the OTP parts are cheaper than the EPROM parts to manufacture thus reducing product costs.

If the production volume is high enough, the EPROM part can be replaced with a mask ROM version. The designer supplies the finished program to the IC manufacturer who creates a mask for the version of the IC that has an internal ROM. This provides the lowest production cost. However, the following caveats exist:

- There is a mask charge to produce the ROM, usually several thousand dollars and usually tied to a minimum purchase requirement. If the product volumes are less than expected or (get your résumé ready) a bug is discovered in the program after the ROM is created, the mask charge is not recoverable. A new NRE (nonrecurring expense) is required for a new mask and all the old parts have to be scrapped, because the ROM program cannot be changed.

- Some manufacturers are so swamped with mask order requests that they have stopped accepting them. This can be disastrous if your entire product pricing strategy is based on the availability of mask ROM parts. A list of these manufacturers, even assuming I knew who they all were this week, would be useless by the time this book reaches print. Check into this before deciding to use a ROM part.

Even though the production costs are low, the high upfront costs prevent many designers from using mask ROM parts. If your volume is too low or you know the design will change before the end of product life, then mask ROMs usually are a poor choice.

One additional consideration is that not all devices are available in all flavors. For example, the Motorola 68HC05 series parts are designed for extremely high-volume applications. Not all parts in the series (and there seem to be more every month) are available in the EPROM version. Some parts are available *only* in the ROM version. Development is done on a similar part for which an EPROM version is available. The catch is, if you cannot justify the ROM costs, you cannot select these ROM-only devices, and the EPROM parts may be too costly for your use.

Another example is the 8031 family parts, which are available in EPROM, OTP, and ROM versions. As of this writing, the cost of the EPROM version is about ten times the cost of the ROM version and the OTP is about 60 percent of the EPROM version, depending, of course, on your volume and where you buy the parts. The basic ROM 8031 may be the cheapest choice, but if you will not have the volume to use it, the OTP version of a different processor may be cheaper than the OTP 8031. The device with the lowest cost in a ROM version may not be the cheapest in the OTP. And, for some devices, the OTP is not available. Your choices are EPROM or ROM, which can make these parts a real cost problem in low-volume applications. Be sure to research which varieties of a part are available based on your volume and other product requirements.

Finally, remember that, once a design is committed to mask ROM, it has the same inflexibility as a non-microprocessor-based hardware design. Once you go to ROM, you give up the flexibility and programmability of having the design in firmware, at least as far as hardware costs go.

In-Circuit Programming

This is not a consideration for every design, but you sometimes need the capability to program the parts in-circuit—to perform field upgrades of the firmware, for instance. This can be a powerful feature, but the capability (or the lack of it) can affect which processor you choose. To use in-circuit programming, you have to have a processor with the program stored in flash memory. You need a way to erase and program the memory without removing the device from the board.

I recently developed a system that needed in-circuit programming. The microcontroller I wanted to use was available in an EPROM version, which has to be erased using UV light. I needed to program the parts without taking them off the boards. There was a flash version of the microcontroller, which could be erased and reprogrammed in-circuit, but it ran at only half the speed of the part I wanted to use. Another version was available with flash memory and capable of running at the right speed, but it had additional, unneeded, features that tripled the cost of the part. I had to compromise on cost or performance, or give up in-circuit programming. Check this carefully if you need that capability.

Nonvolatile Storage

Sometimes your application requires internal nonvolatile storage. If you are building a television, you might want to remember what channel the set last was on, even if power is removed. For this, you will need some kind of nonvolatile storage that can be written and read by the processor. Many microcontrollers, such as the PIC and Atmel AT90S series, include a small amount of EEPROM on-chip.

Memory Architecture

Microprocessor memory architectures are divided into two broad camps: von Neumann and Harvard. The von Neumann architecture permits data and code to be intermixed. You can put a data table in PROM with the code, and you can move code to RAM and execute it there. If the code is in RAM, it can modify itself by writing to the code area of RAM.

The Harvard architecture has separate code and data areas. Code executes from PROM (usually), data comes from a separate RAM, and you cannot get data from the code space. Most microprocessors that use the Harvard architecture, actually use a *modified* Harvard architecture, where the code and data areas are separate but a limited ability exists to get data from the code area. This allows tables or other information to be compiled into the code for use at run time. This usually is implemented with a small number of pointers that can retrieve data from the code space and with the inclusion of immediate instructions where a byte (or word) of data is included in the instruction itself. Many single-chip microcontrollers use the Harvard architecture, among them the 8031, Microchip PIC series, and Atmel AVR90S series. Figure 1.3 shows the relative characteristics of the von Neumann and Harvard architectures.

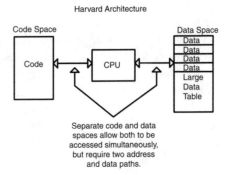

Figure 1.3
Von Neumann versus Harvard Architecture.

The advantage of the Harvard architecture is that there are two separate memory areas and often two separate data paths, so code and data can be fetched simultaneously, increasing the throughput of the processor. From an embedded system point of view, the difference between the architectures is important if compiled data tables are needed. For example, a stepper motor controller may have a number of ramp tables embedded in the code space. If a processor with a modified Harvard architecture is chosen, be sure the table lookup features of the instruction set will not bog down the code. If you are considering an 8031 for this application, you will find that it has several registers that can be used as pointers into the data RAM, but only one register (DPTR) that can be used as a pointer into the code PROM. An application that has to simultaneously use two tables in PROM constantly switches DPTR between the two pointers. One solution to this is to move one or both tables into RAM, but then you must make sure enough additional RAM is available to hold the tables.

Development Environment

To develop applications on a microprocessor, some basic tools are essential:

A development system or cross-compiler

A PROM programmer

Debug hardware

In the prehistoric days of embedded systems (before the IBM PC), the standard development system consisted of a computer from the company that sold the microprocessor ICs and a PROM programmer. The development systems were expensive, slow, and limited to developing software only for that manufacturer's parts. Some third-party companies had development systems as well. These also were expensive and slow but could often be upgraded (at a huge cost) with hardware that would permit development of software for more than one manufacturer's parts. If you can find one of those old development systems today, it probably will be in use as a doorstop or boat anchor.

It is unarguable that the standardization of the business world around the IBM PC and its derivatives has been a real advantage to the embedded systems developer. Most manufacturers of microprocessor ICs now provide development software instead of systems for their parts. These *cross-compilers* run on a PC to compile or assemble code for the manufacturer's microprocessor. (Technically, a cross-assembler is a special case of a cross-compiler, and in this book the term *cross-compiler* will refer to both types of software.) Some manufacturers even give away some sort of development tools (usually an assembler) to potential customers, on the premise that they are in the IC business, not the software business. It is unknown how many microprocessor selections have been made based on the availability of these free tools, but the number has to be large.

Many IC manufacturers still provide complete development systems for their parts. These are usually PCs with the manufacturer's software included, and they constitute a complete development environment. But buy carefully, because they can be a bad deal from a cost perspective.

Some will argue that the PROM programmer no longer is an essential development tool, and they are right. If the project is to be developed in RAM or on an embedded PC or on a flash-based, downloadable processor, a PROM programmer is not needed. As more and more microcontrollers move to a flash-based architecture, the need for PROM programmers in the engineering lab will decline further. However, some projects still are developed in an environment where parts have to be erased and programmed every time the code changes; and for those developers, a PROM programmer is needed.

When the development system consists of a cross-compiler and a PROM programmer and little else, debugging is simple, although often tedious. The code is run, the operation of the system is observed, and the code is examined to see why things do not work. This process is repeated until all the bugs are found. For some systems, especially at small companies, this stare-at-the-code method still is used and works well. This, in fact, was used with the example pool timer system to debug the code. This method becomes less and less attractive as system complexity grows and development schedules shrink.

The next level of debug is a *monitor program*, sometimes called a *debugger*. This simple program resides in the embedded system and provides commands to examine and alter memory, download code, and to insert breakpoints into the code. A *breakpoint* is an unconditional branch instruction that takes the code back to the monitor program, where the registers and memory may be examined. Monitor programs require some kind of terminal, and the monitor program itself takes up considerable memory so they typically are not used with very simple microprocessors.

As microprocessors become more complex, debugging the completed system becomes more difficult. Many designers, especially at large companies, use an *emulator* for system debugging. The emulator has a probe that replaces the microprocessor IC in the system (the *target*) and is supposed to run exactly the same as the target part. However, the emulator allows the engineer to insert breakpoints into the code so that the microprocessor's operation can be stopped. While stopped, the memory, internal registers, and other information about the microprocessor can be examined, the same as with a monitor program. In a simple emulator, the breakpoint typically is a specific address; for example, at the instruction in the pool timer that turns on the motor relay. More sophisticated emulators have additional hardware that allow breakpoints when specific values are written or read to or from memory, when a specific sequence of instructions is executed, or for many other causes.

One drawback to emulators is their cost. Ranging from a few hundred dollars for a simple microprocessor (such as the Intel 8031 family) to several thousand dollars for a more complicated IC, the cost is often prohibitive. As mentioned earlier, many companies base several products on a single micro-processor type due to the cost of buying new emulator equipment.

As microprocessors have grown even faster, their speed has outpaced the ability of the emulator industry to keep up. In addition, the use of more pow-erful processors for applications often means some CPU horsepower is left over after the application is developed. Many developers have moved away from emulators and back to the monitor or debugger programs. These now more sophisticated programs take advantage of leftover CPU capacity to provide event tracing, throughput measurement, code histograms that show how much time the CPU spends in each section of code, and other powerful debugging information. In addition, many processors now include debugging resources on-chip. This will be examined in more detail in a later chapter.

Development Costs

In most companies, someone must produce an estimate of the development cost required for a major product. As for any project, these costs include labor and materials. Estimating these is a matter of experience, which is why it usually is left to the more senior engineers. However, some additional costs must not be forgotten when developing embedded microcontrollers:

Development systems and development software

PROM or other device programmers

ROM mask charges and other NREs

On a large project, these costs usually are minimal. On a small project, they can drive development costs above what the product will produce in sales.

Hardware and Software Requirements

If the product specifications or requirements definition is the goal for the product, the hardware and software requirements are the goal for the detailed design. These requirements start with definition of the user interface and system functionality. In the example system, the complete system definition (see Appendix A) specifies what has to be done and how the user operates the device.

From the system definition, a hardware interface is defined. The most productive method of defining the hardware is to start with the requirements—what the hardware must have. This is tied to the system specifications because the hardware has to support whatever functionality the system has. In the example system, the time must be displayed. Given the constraints of the system (the timer will not be connected to a PC, for example, so a CRT display is out), it came down to two choices: light-emitting diode (LED) and liquid crystal display (LCD). Even though an LCD would be more readable, I chose the LED display because the timer will be out in the weather all year, and the LCD displays available at the time had a problem with cold temperature.

Some people consider each set of specifications to be a fixed, immutable document. I prefer that the hardware specifications be a record of the design decisions. The first section of the hardware specifications is the requirements. This is given to the hardware engineer and becomes the basis for what he or she does. The requirements should spell out just that—the requirements. How the requirements are met is up to the engineer. Anything that cannot be left to the engineer's discretion should be in the requirements document.

When working as a project engineer on a large project, I like to put a list of the requirements for each microprocessor-based board in the engineering specifications. This single document then becomes the foundation for the individual board specifications.

After the requirements document is completed and while the design is progressing, the hardware specifications are updated to include the specific information that another engineer needs to understand the design and that the software engineer needs to program around the hardware. So, when the board specifications are completed, a preface is added that describes the original requirements (and any updates that occurred as the design progressed) and a description of how the design was implemented, with all the information the software engineers need to control the hardware. This includes the following:

Memory and I/O port addresses, including memory maps if appropriate

Amount of memory available

The definition of each bit in each status register

The use of each bit on each port pin

How peripheral devices are driven (such as the clock frequency input to a timer IC)

Anything else the software engineer needs to know about the design

On a complex board, I have often had two separate sections in the hardware specifications. The first describes the hardware and how it works. The second section contains the information the software engineers need to do their job.

In a similar fashion, a software requirements document is created that defines what the software has to do. In a simple design like the pool timer, this may consist of the system requirements document that describes the user interface and the hardware specification that describes how the hardware works.

A detailed software specification that describes the completed design is less common than the equivalent hardware specification. This occurs for three reasons: First, the hardware specification is passed to the software engineers so they will know how to manipulate the hardware. Usually no corresponding "customer" needs to know the technical details of the software, so the need for documentation is not as great. Second, software is easy to change so it changes frequently, often whenever someone in marketing thinks up a new feature to add. In some situations, software specifications can be very hard to keep up to date, especially if the software engineers have other, higher priorities. The last reason for a lack of finished software specifications is that software usually is the last part of a project to be finished and often not enough time is left at the end of a project to document it.

That said, I will point out that company or customer policy sometimes requires detailed software specifications. For example, defense projects usually require extensive documentation detailing every function that the software performs.

In a simple design, the software definition, like the hardware definition, may describe the software for a single board. In a more complex design, where different software engineers work on different parts of the code for a single board, there may be a software definition for each individual engineer's code. In a complex multiprocessor system, there may be an overall software document, which I consider to be part of the system engineering specification.

The software specifications should include the following:

- A statement of the requirements, including the requirements definition, engineering specifications, and hardware definition, as appropriate.
- The communication protocol to any other software, whether to another processor or to another piece of the software for this processor. This should include descriptions of buffer interface mechanisms, command/response protocols, semaphore definitions, and in short, anything to which the complementing code needs to talk.
- A description of how the design was implemented, using flowcharts, pseudocode, or other methods. (Chapter 3 describes these in more detail.)

Since software can be broken down more flexibly than hardware, it is difficult to pin down a single software definition format that works for everybody all the time. The key is to define any interfaces that other engineers need to know about and identify the design details that engineers in the future might need to know.

This discussion assumes that the hardware and software are fairly independent. In a simple system like the pool timer, that is a good model. The hardware is designed, the software is written around that hardware, and that is that. While the actual design implementations may proceed in parallel, the software engineer basically writes code around the hardware available.

In a more complex system, the process may be iterative. For example, the software and hardware engineers may have a meeting where they jointly decide what hardware is required to perform the function. Large amounts of memory may be required for data buffers, or the software group may request a specific peripheral IC because an interface library already has been developed for it. There are trade-offs, in this game, between ease of software development and cost or complexity of hardware.

Hardware/Software Partitioning

Once, while having lunch with a group of engineers, I jokingly made the statement that my design philosophy was to put everything under software control. That way, bugs in the design by definition were the fault of the software engineer.

This flippant conversation touches on a real problem in any embedded system: Which functions should be performed in hardware and which in software? An example of this can be found in the pool timer. As we will see in the next chapter, the pool timer displays time information on four seven-segment LED displays. Display decoder ICs accept a four-bit input and produce the signals necessary to drive the display. This design takes a different approach and drives the display segments directly from a register, which is under software control. When the software wants to display a number, it has to convert the number to the seven-segment pattern and write that pattern to a register. The savings was a single IC in the design. This approach also allows the code to display nonnumeric symbols on the display (A, C, H, J, L, P, U), which I used for debugging the system.

While this decision saved an IC, it had three costs: ROM space was needed for the lookup table, extra code had to be included for the hex to seven-segment conversion, and the software needed extra time to perform the translation. Given the simplicity of this function, none of these was a serious problem. The table was 16 bytes long, the code took a few more bytes and needed only a few microseconds to execute. But the principle described is at the heart of the software/hardware trade-off. The more functions that can be pushed into software, the lower will be the product cost, up to the point where a faster processor or more memory is required to implement the added functionality.

The pool timer demonstrates another example of this kind of trade-off, which is discussed in Chapter 4.

As the saying goes, there is no such thing as a free lunch. Pushing functionality into the software increases software complexity, development time, and debug time. This is an NRE, just like the mask ROM charges described earlier. However, given the increasing speed and power of microprocessors, I expect to see an ever-increasing trend toward including as much functionality in the software as possible.

In a more complex system, these trade-offs can create heated discussion. Should the software handle regular timer interrupts at a high rate and count them to time low-rate events or should an external timer be added that can be programmed to interrupt the software when it times out? Should the software drive the stepper motor directly or should an external stepper controller be used? If the software drives the motor, should protection logic be included to prevent damage to the motor drive transistors if the software turns on the wrong pair? And if the processor runs out of throughput halfway though the project, did the design place too much burden on the software or did the software engineer write inefficient code? The answers to these questions depend on your design. But if you stay in this field very long, be prepared to get into one of these discussions.

While doing everything in software increases development costs, moving functionality to the hardware increases product cost, and these costs are incurred with every unit built. In a low-cost design, addition of any extra hardware can have a significant effect on product cost, so the software/hardware trade-off can be extremely important. In an extremely cost-sensitive design, such as a low-cost consumer product, functions that cannot be performed in software may simply be left out.

Distributed Processor Systems

Multiprocessor systems will be covered in more detail in Chapter 7. Here, we summarize the trade-offs involved in choosing a multiprocessor architecture.

A distributed processor system might have a single CPU that communicates with a host computer and distributes commands and data to lower-level processors that control motors or collect data from sensors or perform some other simpler task.

Distributed processor systems have the following advantages:

- The actual processing hardware can be located near the device being controlled or monitored. In large equipment, this may be a real advantage.
- If some of the functionality is optional, the cost of the processor that controls the option can be added or removed with the option.
- In a distributed processor system, each of the distributed CPUs usually can be a lower-performance (cheaper) part than would be required for one central CPU.
- A distributed system can be designed with a better match between the CPU and the task it must perform. In a single-CPU system, the CPU must be fast enough, have enough memory, and so forth to perform all the tasks, whether simple or complex.
- The code for any given CPU in a distributed system usually is simpler.
- It is easier to determine if the CPU power is adequate in a distributed system, because fewer tasks are being swapped in and out and there is less interaction among the various processing that has to be performed. For example, you need not worry about how the motor control function affects the serial interface throughput if the tasks are handled by separate processors.
- Debug of distributed systems can be simpler, since each processor performs a limited set of tasks.

The advantages of a single-CPU system are:

- Synchronization, when needed, is easier. For instance, it is easier for a single-CPU system to synchronize motor start-up to limit current surge, simply by communication between tasks or by scheduling. In a distributed system, such synchronization must be performed by CPU-to-CPU communication or back through a common control CPU.
- All the data is in the same place, making communication with a host or other systems easier. Fewer communication protocols are required to pass data around.
- Since there are fewer oscillators, there usually will be less EMI. On the other hand, a faster processor may be required, operating at a higher frequency and generating a lot of EMI.
- If the design changes so that intertask communication must be added, such as for motor synchronization, a distributed design may require that interfaces be added to each distributed CPU. In a single-CPU design, such a change is likely to be only to the software.
- It is easier to download or update code in a single-CPU system.
- Debug of a single-CPU system may be easier since all the functions are in a single place and all the interactions can be examined. Of course, these interactions, task switching, and general complexity of the code can complicate debug as well.
- Fewer development tools are needed, since there is only one processor. In a distributed system, the same thing can be achieved by using only one type of CPU, but that defeats the ability to match the CPU to the task.
- If an RTOS is used, there will be fewer license fees in a single-CPU system. But a more complex RTOS may be required.

With increasing processor power at decreasing cost, I think more single-CPU designs are to be expected. Some designs will take advantage of increased CPU horsepower to add new functions, such as real-time signal processing. But motors and other electromechanical devices are getting no faster, so systems that interact with these devices probably will use fewer, more powerful processors. Complex systems that use a single Pentium-class CPU and a few 8-bit microcontrollers as smart sensors would not be surprising.

Specifications Summary

Let us summarize the contents of the design documents described in this chapter before we look at the actual design of the rest of the book.

The requirements document describes:

What the design or system is to do

The user interface, if any

Any external interfaces to other systems

What the real world I/O consists of

Hardware specifications (one per board or subsystem) describe:

The requirements, restated from engineering or requirements documents

How the hardware implements the functionality

The software interfaces to the hardware

Software specifications describe:

The requirements

Interfaces to other software

How the software implements the requirements

A Requirements Document Outline

The following is an outline for a requirements document that will fit most products. This document describes the product as a "black box"—what the product does, not how it is done.

Overview. A brief description of the document, such as "This document describes the requirements for the ABC corporation swimming pool timer."

Related documents. Related internal documents, such as product specifications, environmental specifications, and the like. Related industry specifications such as ANSI or IEEE specifications.

Specifications. These could include the following:

Agency approvals. List agency approvals that the product must meet, such as FDA requirements, IEC 950, UL 1950, shock/vibration specifications, and so forth.

Requirements. List system requirements. The following items are typical of the sort of thing that might be listed, and obviously all of these items will not apply to all products. This section is the core of the document and may run to dozens of pages.

MTBF (mean time between failure)

MTTR (mean time to repair, usually applies to products that are serviced by a field service organization)

Speed (How many things per minute/hour/day must be done?)

Operator interface (LCD? touch panel? barcode readers? mouse/keypad?)

External interfaces (interfaces to other systems, to a controlling host system, or to a slave subsystem. Ethernet? RS-232? Proprietary?)

Available options (may be lengthy if several need to be described)

Input power (list input voltages, frequencies, and current; include international requirements)

Export restrictions and requirements (applies if using controlled technology; also, requirements for the product to be marketed in certain countries may limit technology that can be used)

Input requirements (What size bottles does it use? What sizes of paper can it handle? How big or how small can the block of steel be that goes into the input hopper?)

Capacity (How many blocks of steel or bottles or pieces of paper can it handle at a time?)

Error handling (What happens if the operator puts in too many bottles or a block of steel that is too heavy? What happens if power goes off halfway through the process?)

Weight (usually applies only to large or portable products)

Size (Does it have to fit through a standard door or on a standard elevator? In a standard briefcase?)

Safety requirements: (Does it have to operate in standing water with no danger of electrocution? Does it need a safety mat to stop the robotic arm when a person steps inside the fence? Are there rotating mechanisms that must be covered or stopped when a door is opened? Must the operator be protected from high temperatures?)

External interfaces (interfaces to external systems, like a 100 base-T Ethernet interface to a computer network or an IRDA interface to transfer data to and from a PC)

Note that there may be other requirements as well, such as media requirements, customer versus field engineer maintenance items, and the like. However, since we are concentrating on embedded systems, those are outside the scope of this outline.

Hardware Design

Once the system is designed and the hardware requirements are established, the next step is to design the actual hardware. Of course, you will document the design to make life easy for the software engineers, right?

Embedded microprocessors fall into two categories: Single-chip embedded solutions with on-chip memory like the 8031 and embedded systems using a microprocessor with external memory and I/O. Examples of these are a 68000-, 80186-, or 386EX-based embedded system.

Figure 2.1 shows the simplest single-chip microprocessor designs and multichip designs. Note that they are basically the same, except that the single-chip design has everything inside the chip (inside the dashed line) and the multichip design has everything except the processor itself outside.

Single-Chip Designs

Single-chip microprocessors (or microcontrollers) usually provide erasable programmable read-only memory (EPROM; or ROM or flash memory), random access memory (RAM), and I/O ports. Most also have internal timers, serial interfaces, or other peripherals. The I/O ports are flexible, permitting each bit to be assigned as input or output.

The actual design of single-IC systems is straightforward. Before starting the design, you already know (or should know) that there are sufficient I/O port pins, enough internal memory, and sufficient processor speed to do the job.

A single-IC design requires an external timebase. This can be a clock from some master source (such as a higher-level control processor), a crystal, a ceramic resonator, or even a resistor/capacitor timing circuit on some processors. What you use depends on your cost requirements and how accurate the timebase needs to be. If you are using a crystal or resonator, connect

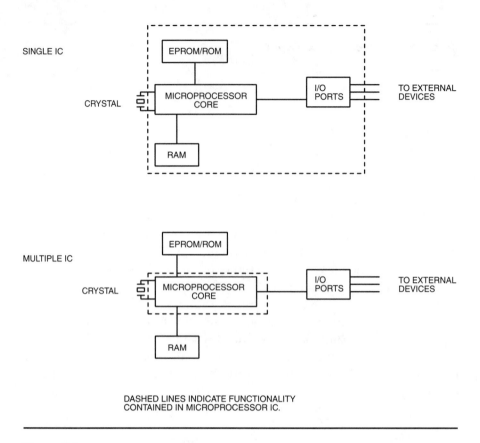

Figure 2.1
Single and Multiple IC Microprocessor Circuits.

it according to the manufacturer's data sheets. If you are using an external clock source, such as a packaged oscillator, make sure it meets the voltage and capacitance drive specifications of the processor.

Multichip Designs

While the similarity between single-chip and multichip designs shown in Figure 2.1 is correct, it is somewhat misleading. The architectures are similar, but in the real world, a multichip design usually is more complex. There usually is more memory and generally more (or more complicated) I/O ports. A single-chip microcontroller may not be suitable for a design for many reasons: insufficient I/O pins, insufficient RAM or ROM, or any of the other considerations detailed in Chapter 1. However, once a decision has been made to go to a multichip implementation, you take a quantum step in complexity. A multichip design usually has most or all of the following as separate components:

Microprocessor

Random access memory (RAM)

Programmable read-only memory (PROM)

Peripherals (I/O devices)

The following table illustrates typical memory configurations for various microprocessors. The Atmel part is an 8-bit microcontroller, the NEC part is a 32-bit microcontroller, the 80C188 is a midrange microprocessor, and the Pentium is a high-end microprocessor.

Processor	PROM/ROM	RAM
Atmel AVR90S4414 (internal RAM/ROM)	4K	256 bytes
NEC V853 (Internal RAM/ROM)	128K	4K
Intel 80C188	512K	512K
Intel Pentium	4MB	4MB

Compared to a single-chip design, a multichip design costs more, takes more PC board real estate, and is more complicated. The benefits are more flexibility, more expandability, and (usually) more processing power.

In a multichip design, external peripherals (timers, I/O ports, analog-to-digital converters, etc.) must be connected to the microprocessor using the data and address buses. There are several types of microprocessor bus cycles, but all do the same basic things: The microprocessor generates an address, which is decoded to select a peripheral (memory or I/O) device. If the cycle is a read cycle, the processor supplies a signal to tell the peripheral to drive its data onto the tristate data bus for the processor to capture. If the cycle is a write cycle, the processor drives the write data onto the data bus and generates a signal indicating that the peripheral should capture the data.

Figure 2.2 shows typical timing diagrams for four families of processors: Intel, Microchip, Zilog, and Motorola. The speed of the signals varies greatly from one processor to the next but the basic waveform is the same for processors within a given family.

The Intel timing diagram applies to Intel processors from the 8085 to the 80188/80186. It also includes the Intel microcontroller 8x3x/8x5x family of parts, when those devices are used with external memory. Other

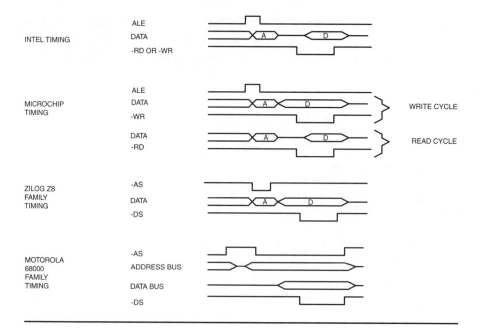

Figure 2.2
Typical Microprocessor Bus Timing.

manufacturers, such as Philips, also make variations on the 8x31 family that use Intel-type timing to access external memory. The NEC µPD7840xx micro-controller family uses Intel timing, as does the Siemens/Infineon C167 family when accessing external memory in a multiplexed mode.

In the Intel scheme, the data bus is multiplexed with the address bus. In a processor with an 8-bit data bus, the 8 data bits are multiplexed with the 8 lower address bits. If the data bus is 16 bits, then all 16 data bits are multiplexed with the lower 16 address bits. Multiplexing is a common means to access external memory because it saves pins—without multiplexing, accessing 64 kilobytes (K) of memory with an 8-bit bus would require 24 pins just for the address and data lines. A multiplexed scheme requires only 16 pins for address and data.

During the first part of the machine cycle (labeled *A* on the diagram), the microprocessor places the address on the data bus, and it must be captured by an external latch such as a 74AC373. The ALE (address latch enable) signal is used to capture the address in the external latch. After ALE goes inactive, the processor generates a read or write strobe (-RD or -WR) to transfer data to or from the external memory or I/O device. For a read cycle, -RD is driven low, indicating to the peripheral device that it should drive read data onto the bus, which the processor will leave in the tristated condition. For a write cycle, -WR indicates that write data is available for the peripheral, and the proces-

Embedded Microprocessor Systems

sor will drive the data onto the data bus. This basic waveform is used whether the external device is an EPROM, RAM, or peripheral.

The second waveform in Figure 2.2 shows the timing for external memory access by a Microchip PIC17Cxx part. The basic waveform is nearly identical to the Intel, with one significant difference: During a write cycle, the Microchip part places write data on the data bus prior to the leading (falling) edge of the -WR strobe. With the Intel timing, write data is guaranteed to be stable only prior to the *trailing* (rising) edge of the -WR strobe. Other devices that use this same basic timing are the Atmel AT90S4414 and AT90S8515 microcontrollers, when accessing external RAM.

The third waveform shows the timing used by the parts in the Zilog Z8 family. The data bus is still multiplexed with the address, but the address strobe (-AS) is true when low, instead of when high. There are no separate strobes for read and write. Instead, there is a single data strobe (-DS) and another signal (R/W) that determines if the cycle is a read or write cycle.

The fourth waveform in Figure 2.2 shows timing for processors such as the Motorola 68000 family. These parts have separate address and data buses. The address strobe is not used to latch the address but to indicate that a valid address is present on the bus. Similarly, the data strobe is used to indicate that valid write data is present on the data bus (write cycle) or that the peripheral should place read data on the bus (read cycle). The 68000 family parts also use a -DTACK (data transfer acknowledge) signal from the addressed device to indicate the end of the data transfer cycle.

The timing sequences shown in Figure 2.2 cover the majority of microprocessors and microcontrollers that can access external memory. There are some other memory access schemes as well. The Siemens/Infineon C167 family, mentioned earlier, has a multiplexed mode that follows the Intel timing. The C167 parts also have a demultiplexed mode that eliminates the external address latch. Since the address is demultiplexed inside the chip, this mode requires an additional 16 pins for the address signals. The ALE signal still is generated to indicate a valid address, but external address latches are not required.

The Intel 80C960 family is a 32-bit embedded processor that also uses a demultiplexed address and data bus. Like the 68000, the 80C960 uses a low-true address strobe to indicate a valid address.

The Zilog Z180 and Z380 microprocessors, not shown in the figure, use timing similar to the Intel timing, with separate read and write strobes. However, these parts do not multiplex the address lines with the data lines, so there is no need for an ALE signal to latch the address. There are dedicated address pins on the part, and the address is stable throughout the bus cycle. A separate -IORQ or -MREQ line goes active to indicate whether the bus cycle is a memory or I/O operation.

The Z380 also provides an indication, similar to the ALE signal, when a bus cycle starts, for designs requiring that information.

Figure 2.3 shows how a 74AC373 latch would be used to capture the multiplexed address on one of the processors that uses a multiplexed address/data bus. The address is latched so that when the multiplexed bus switches to data, the address is still available for the peripherals to use. The circuit shown in Figure 2.3 is typical of a processor with an 8-bit external interface. When using a processor with a 16-bit data bus (such as the Intel 80186), both bytes of the bus are used for data transfer, so two 8-bit latches are required to capture the full 16-bit address bus.

The output enable signal to the 74AC373 is shown grounded. This enables the outputs, and therefore the address bus, all the time. There are some circumstances where this will not be the case, which will be covered later.

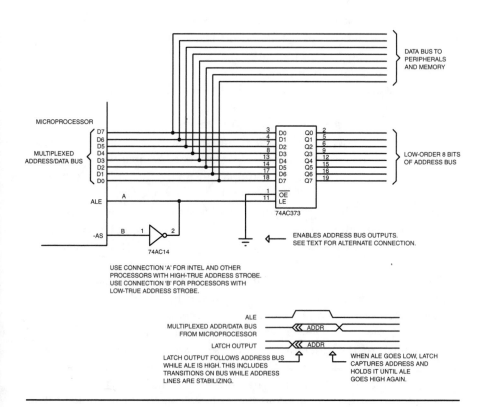

Figure 2.3
Address Bus Demultiplexing.

The latching circuit need not be a duplicate of the one shown in Figure 2.3. It could be implemented in a programmable logic device (PLD) or other logic.

One final note: So far, we have discussed only 16-bit address buses, which allow access to 64 K of memory. Many processors can address more than this. In some of these parts, including the 80188/186 family, an additional latch (or latches) is required to capture the upper bits of the address if it is needed, since it is multiplexed with some status signals.

Wait States

In many cases, a fast microprocessor must interface with a much slower peripheral. In this case, the normal timing of the microprocessor read, write, or data strobes is much too fast for the peripheral. For example, the processor may generate a -RD signal that is 200 ns in length, but the peripheral has a 300 ns output enable time. In these cases, the usual solution is to add *wait states* to the bus cycles that access that peripheral. A wait state extends the microprocessor read or write cycle by an integral number of processor clock cycles.

Not all microprocessors support wait states; for example, most single-chip processors (e.g., 8051, PIC17C4x) do not have a provision for wait states. However, most processors designed for multichip applications support wait states.

Internal Wait States

Some processors have internal logic that can insert wait states. These are programmed in software to extend processor cycles when accessing specific memory or I/O addresses. The 80186 has several outputs that can be programmed to generate chip selects at specific address ranges. These can be used to select EPROM, RAM, or I/O devices. For each output, an internal wait state generator can be programmed to automatically insert up to three wait states. They also can be programmed to either accept or ignore wait requests from the external wait signals.

Wait State Timing

When the processor starts a bus cycle and detects the wait line active, it will extend the cycle, leaving the -RD, -WR, or -DS signal active and sampling the wait line once per clock. Once the wait signal has gone inactive, indicating that the peripheral is ready, the processor will complete the bus cycle.

The wait input is conceptually straightforward, but the details can cause problems. The most common problem is timing assertion of the wait state, which requires study of the data sheets. Figure 2.4 shows a (simplified)

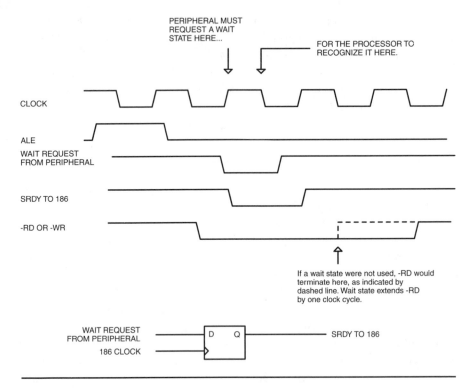

PERIPHERAL MUST
REQUEST A WAIT
STATE HERE...

FOR THE PROCESSOR TO
RECOGNIZE IT HERE.

CLOCK

ALE

WAIT REQUEST
FROM PERIPHERAL

SRDY TO 186

-RD OR -WR

If a wait state were not used, -RD would
terminate here, as indicated by
dashed line. Wait state extends -RD
by one clock cycle.

WAIT REQUEST
FROM PERIPHERAL

186 CLOCK

D   Q

SRDY TO 186

Figure 2.4
80186/80188 Wait State Timing.

diagram of the 80186/80188 processor timing. The SRDY (Synchronous ReaDY) input of the 80186 has to be asserted prior to the second falling clock edge after the ALE goes inactive. But SRDY has to be externally synchronized by the user, so the peripheral actually has to assert the wait request right after the -RD or -WR signal. If the wait logic is delayed too much, the request will occur too late, and the processor will ignore it. Other processors have different quirks, which must be taken into account.

Some peripheral ICs include integral wait state generators. If you use one of these, be sure that the timing will work with the processor. Some peripheral ICs assert the wait request too late in the cycle for some processors to recognize it.

Bus Types and Their Relationship to Wait States

Processors like the Intel x86 family use a synchronous bus. They do not use a bus acknowledge signal and they default to no wait states. In other words, the input (usually READY) that causes wait states to be inserted in the cycle normally is pulled to the ready (no wait state) condition. If the external logic does not drive the input to generate wait states, the processor generates the access cycle and continues on, whether the peripheral was really ready or not.

Processors like the Motorola 68000 family use an asynchronous bus. In this scheme, each peripheral must return an ACK signal to indicate that it has completed the data transfer (accepted the write data or generated the read data). Asynchronous timing means that the default operation of the processor is to wait until the peripheral responds, which may be forever if the peripheral does not acknowledge the transfer. In theory, access to nonexistent memory or a nonresponding peripheral will cause a permanent wait state. In practical systems, a timeout circuit usually generates an ACK (or more specifically, an error signal) if the peripheral does not.

In a synchronous transfer, a peripheral needing wait states must detect when it is being accessed and drive the processor's ready input to the inactive (not ready) state until the peripheral has had time to complete the read or write operation. The ready input then is driven active, permitting the processor to complete the cycle.

In asynchronous systems the peripheral must generate an ACK to indicate that the transfer is complete. In actual systems, the peripheral itself usually does not introduce the wait states. This normally is done by the logic that controls access to the peripheral device, which times wait or ACK assertions and makes sure that they are asserted only when the correct peripheral is accessed. Some peripherals (particularly those designed for the 68000 family) generate ACK internally and need no external logic for this function.

DMA

DMA (direct memory access) is a means of having two or more processors share the same bus. When a secondary processor wants control of the bus, it notifies the first processor, which gives up the bus. The second processor then drives the address, data, and control lines and accesses a peripheral just like the first processor. Typical examples of DMA uses are to permit two processors to share common memory, to refresh dynamic RAM, or to transfer data from an I/O device (such as a serial port) directly to processor RAM.

Processors that support DMA provide one or more inputs that the bus requester can assert to gain control of the bus and one or more outputs that the processor asserts to indicate it has relinquished the bus. When designing with DMA, address buffers must be disabled so the bus requester can drive them without bus contention. On the 80188, for example, the HLDA (HoLD Acknowledge) signal indicates that the processor is acknowledging a

DMA request. It can be connected to the address latch output enable pins, which will tristate the outputs when the processor is in a hold state. If data bus buffers are used, a similar mechanism is needed to disable them. DMA can be used to implement a multiprocessor communication scheme, which will be described in the chapter on multiprocessor systems.

DMA Controllers

In a DMA scheme, the second processor may not be a processor but a DMA controller. This peripheral device takes control of the bus but does no actual processing of instructions. Instead, the DMA controller is used to move data between another peripheral device and the microprocessor's memory. A DMA controller contains counters that automatically increment to the next address after each transfer, so blocks of memory can be moved. An example DMA controller would be the one in your PC that moves data from the hard disk controller into memory. DMA controllers permit the microprocessor to be performing some other operation while a data transfer happens in the background. The microprocessor just sets up the DMA and processes the entire block of data when the transfer is complete. Figure 2.5 shows the basic timing of the DMA process.

Memory

The reason we need to have the address bus captured in a latch is because EPROMs, RAM, and most other peripheral devices need a stable address input during the external bus cycle. Figure 2.6 shows an EPROM connected to a microprocessor with a multiplexed address bus. This example shows an 8-bit EPROM connected to a microprocessor with an 8-bit data bus. If the microprocessor had 16 data bits, the upper 8 address bits would come from a latch as well, instead of directly from the microprocessor. Of course, a 16-bit EPROM would connect to all 16 bits of the data bus.

Types of PROM

Three types of memory ordinarily are used as PROMs in embedded systems. The first of these is the EPROM. An EPROM consists of an array of transistors that can be programmed. The code to be executed is programmed into the device, and it is read out by the microprocessor. EPROMs have a quartz

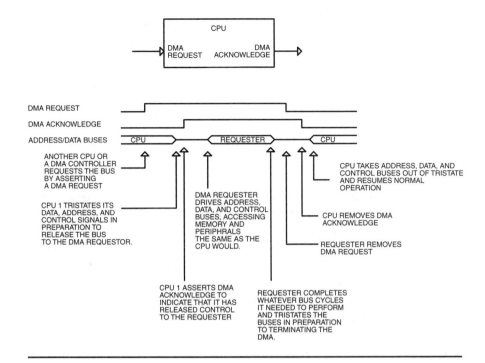

Figure 2.5
DMA Operation.

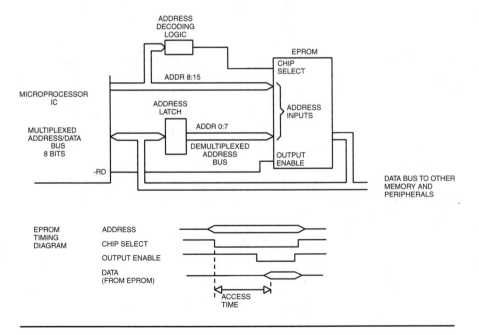

Figure 2.6
EPROM Connected to a Microprocessor with a Multiplexed Address/Data Bus.

window in the top through which the IC die can be seen. This allows the EPROM to be erased using ultraviolet light and then reprogrammed.

One special case of EPROMs is OTP (one-time programmable) PROMs. As mentioned in Chapter 1, these are EPROMs in a plastic package with the quartz window missing. They can be programmed once, but because there is no erasure window, they cannot be erased and reprogrammed. EPROMs and OTP PROMs are programmed using a tool called a *PROM programmer*. EPROMs and OTP PROMs can be either part of a single-chip microcontroller IC or general-purpose parts for use with any multichip microprocessor design.

Another type of memory is flash memory. *Flash memory* is similar to the EPROM in that a transistor array is programmed. However, flash memory can be erased electrically, which means it can be reprogrammed without taking it out of the microprocessor circuit. Flash memory often is used when the product requires that the firmware be upgraded in the field. Early flash memories were expensive compared to EPROM, but the pricing is such that most new designs are flash based.

The advantage of flash memory is that it can be programmed in-circuit, usually by the microprocessor that uses it. The programming procedure requires that the memory first be erased. This can present a problem—if the code to program the flash memory resides in the flash memory itself, how do you reprogram it? This often was a real problem for designers using early flash memories. One way to fix the problem is to move the programming code into RAM and execute it from there. Another approach is to use a newer block type of flash memory. These devices do not require that the entire memory be erased, instead permitting the memory to be erased in blocks. So the programming code can reside in a section of memory that is not erased, while the code resides in another part of memory that is erased and reprogrammed as needed. The Atmel AT49F080 is a 1 MB × 8 flash memory with a 16 K boot block. Two versions are available, one with the 16 K boot block at the bottom of the memory (starting at 00000) and one with the boot block at the top of memory (starting at FC000). Normally, you would put your initialization code and flash erase/programming code in the boot block. This allows you to reprogram the rest of the memory to update the firmware.

Programming flash memories typically requires a specified sequence of writes to specific locations. The Atmel AT49F080 uses the following sequence to initiate the erase cycle:

```
Addr   Data
5555   AA
2AAA   55
5555   80
5555   AA
```

```
2AAA   55
5555   10
```

Once erase is started, the memory will complete the operation itself, timing it internally. Similar command sequences are used to protect and unprotect the boot block, to request the manufacturer's ID from the device, and to program a byte in the memory.

Most flash devices use -DATA polling when programming. This allows the processor to poll the device by attempting to read the location just programmed. The flash memory returns the complement of the data that was written, until the internally timed programming cycle is complete.

Not all block-organized flash memories have a single small boot block and a larger main block. Some have multiple boot blocks, and some divide the memory into a few large blocks.

Most modern flash devices can be programmed using only the normal supply voltage (5V, 2.7V, or 3V). Internal charge pumps generate the higher voltage (typically 12V) needed for programming. Flash devices also can be programmed in a PROM programmer, which usually allows the boot block erase lockout to be overridden. Some microcontrollers with internal flash memory require an external programming voltage.

When not being programmed, flash memories work like EPROMs. The flash memory will have an additional input that controls writing of the memory array.

Flash memory devices also have a means to read the device manufacturer and ID code. This is useful for device programmers, but it also often is needed for in-circuit programming. Different manufacturers have different algorithms for erasing and programming flash memory. If you want to have multiple sources for the flash memory in your design, then your software will need to read the flash to determine which device is installed, so it can determine which programming algorithm to use. You also will need to retain multiple programming algorithms in memory, one for each type of device you can substitute into the system. This was no problem with EPROMs—all 27256 EPROMs work the same when reading. Programming differences were taken care of by the device programmer. On the other hand, EPROMs cannot be reprogrammed in-circuit.

Flash memory is a type of electrically erasable PROM (EEPROM). For general use in program storage, devices designated as EEPROM mostly have been replaced by flash memory. However, in some applications specialized EEPROMs are very common. These will be addressed later.

When an EPROM needed to be erased and reprogrammed, you just pulled it out of the socket and took it to a PROM eraser and then to a programmer. As flash memories have grown in density, this becomes impractical, since they no longer fit in a dual inline package (DIP). Early flash parts were available in PLCC (plastic leaded chip carrier) packages, which could be socketed, but many newer parts are only available in packages such as TSOP (thin small outline package) or BGA (Ball grid array) that are difficult or impossible to put in a socket. The parts are soldered on the board. The result is that, in many designs, the only way to program the flash memory is in-circuit, using the microprocessor itself. This is fine if you want to program a flash memory to update an existing program. But how does the program get into the memory in the first place?

Flash memories still can be programmed by a programmer, using a special socket, before they are installed on the board. But what happens if the vendor (or your own Manufacturing department) inadvertently skips the program-ming step? Or what if you get a batch of boards with the wrong program in the flash? Do you scrap the entire lot?

Some designs are intended to have the flash memory programmed when the boards are tested, by an in-circuit programmer. This is common in high-volume designs. In a design that does not use this technique, it is a good idea to provide a means to program the flash using an external fixture. To do this, the microprocessor must tristate its address and data lines so the external circuit can get to the flash. If the microprocessor supports DMA, it can be put in a hold state. Many processors will tristate their buses if they are held in reset. If buffers or latches are used for the signals or if the signals pass through a PLD before they reach the flash, they can be tristated there.

Once the processor has released the bus, some means must be provided to access the flash memory. This can be accomplished with a connector that brings out the address/data buses and control signals. If there is no room for that, a matrix of pads on the PCB, accessed with spring-loaded test pins, can be used instead.

Finally, an alternative to directly programming the flash is to provide a means, such as a header, to install a daughterboard containing a small flash memory that replaces the system flash. By mapping part of the system flash to a different location in memory when the daughterboard is installed, the boot portion of the system flash can be reprogrammed. Such remapping can be accomplished with a jumper on the main board, or it could be automati-cally activated when the daughterboard is installed. The boot portion of the system flash, of course, would be programmed with code that permits the remaining flash memory to be programmed.

The last type of memory is ROM. As mentioned in Chapter 1, this is memory programmed by the integrated circuit (IC) manufacturer using a mask. It cannot be reprogrammed and usually is used in single-chip micro-controllers, although mask ROM versions of some EPROMs are available.

ROM normally is used only in very high-volume applications, where the code is not expected to change over the life of the product.

Typical EPROMs have three inputs: The address inputs, which can be up to 18 bits; a chip select; and an output enable. The only outputs are the 8 or 16 data bits back to the microprocessor. Figure 2.6 illustrates the EPROM timing diagram. The address is presented to the EPROM and the chip select is driven low. Until the access time has elapsed, the output data is undefined. After the access time has elapsed, the output data for the addressed location is available. The output enable signal turns on the tristate EPROM outputs, driving the data onto the microprocessor data bus.

The chip select signal comes from the address decoding logic connected to the microprocessor data bus. Some processors, such as the 80188/80186 family, have internal, programmable chip select logic. The chip select signal in those cases can come directly from the microprocessor itself.

Note that the address must be stable for the entire EPROM access cycle. If the address changes during the cycle, the outputs also change as the EPROM attempts to access the data at the new address.

The access time of the EPROM is a critical factor and is often overlooked in embedded designs. EPROM access times are specified as a maximum. For example, an EPROM with a specified maximum access time of 120 nanoseconds (ns) requires no more than 120 ns from the time the address is stable and chip select is low to generate a valid output. Most of these EPROMs will be faster than the maximum time specified, which gets a lot of designers into trouble. If you do not take into account the worst-case numbers, the design will work until the Purchasing department buys a batch of EPROMs that happen to be a little slower than the ones you used in engineering debug. Worse yet, the problem may show up only when the temperature is above 90°F or when a certain brand of microprocessor is used.

Calculating EPROM Access Time

To calculate the required EPROM access time, you have to start with the microprocessor data sheets. The procedure is as follows:

Calculate the time from when the microprocessor provides a stable address until it requires stable data.

Subtract any delays, such as the address latch propagation delay.

The result is the required EPROM access time. Any EPROM with an access time at least as fast as the calculated number will work.

Figure 2.7 shows a simplified, typical timing diagram for a processor with a multiplexed data bus. There are three clock cycles from when the microprocessor outputs the address until it requires stable data. However, due to internal delays

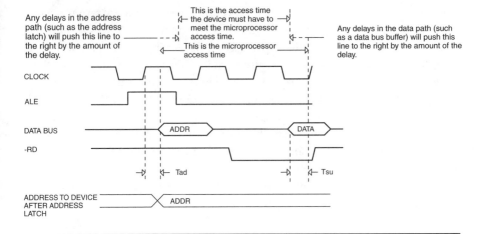

Figure 2.7
Typical Microprocessor Timing.

in the microprocessor IC itself, the address is not available until some time (*Tad* on the diagram) after the first clock edge. Then, the processor needs the EPROM data stable some time before the clock edge that captures the data, because the internal data latch has a finite setup time. This is time *Tsu* on the diagram. The address has to propagate through the address latch before reaching the EPROM, so the latch propagation delay time must be added to the EPROM access time. The required EPROM access time then is

$$3 \times \text{clock period} - \text{Tad} - \text{Tsu} - \text{latch propagation delay}$$

Some microprocessor data sheets make this easier by referencing everything to the control signals (ALE, -RD, etc.) themselves.

Figure 2.7 shows the delay that the address latch causes in the address signals as well as the delay that would be introduced by a data buffer between the EPROM and the microprocessor. You can see that the effect of any propagation delay in the address or data path is to shorten the available access time by the sum of all the delays.

A processor with a nonmultiplexed data bus will have different timing from that shown in Figure 2.7, but the basic concepts are the same. The processor will assert the address some delay after a clock edge, a control strobe will generate some delay after another clock edge, and the processor will want data to be stable on the rising edge of the control strobe or on the clock edge preceding it. The EPROM must be fast enough to produce data in the time from when the address is stable to when the processor needs the data, minus any delays in the data or address paths.

Embedded Microprocessor Systems

For most EPROMs, the access time from chip select is the same, or nearly the same, as the access time from the address. Referring again to Figure 2.7, the EPROM chip select is generated by address decoding logic. The procedure for calculating the chip select access time is the same as for the address access time, except that the delay through the address decode logic must be subtracted from the total time available. If the upper address bits are latched and then decoded to generate the chip select, *both* the latch delay and the decoder delay must be subtracted from the total time. After the address and chip select access times are calculated, the EPROM speed required is the *smaller* of the two numbers.

The next EPROM parameter is the output enable time. This is the time from when the microprocessor asserts the -RD strobe (or the equivalent signal) to when it needs stable data available. In most cases, an EPROM selected to meet the address/chip select access time will not cause a problem with the output enable time. However, it should be checked. Calculating the output enable time is similar to calculating the access time:

Calculate the time from when the microprocessor asserts the -RD signal until it requires stable data.

Subtract any delays, such as the data bus transceivers.

The result is the required EPROM output enable, which can be expressed in equation form like this:

$$Toee = Toem - Td$$

where Toee is the required ouput enable time of the EPROM; Toem is the time from when the microprocessor asserts -RD until it needs stable data; Td is the sum of any circuit delays, such as gating logic in the -RD signal or data bus buffers.

The last parameter is the EPROM data hold time. This is the time from when the output enable (OE) signal goes high until the EPROM actually stops driving (tristating) its pins, sometimes called the *data bus release time*. This time is important because, if the EPROM is still driving the data bus when the processor starts the next cycle, there will be bus contention and the wrong address can be latched. In most cases, selecting an EPROM that is fast enough for the processor also results in the data hold time being fast enough. But when a very fast processor is interfaced to a slow EPROM, the hold time can be a problem. If the calculated hold time is a problem, the solutions are to use a data buffer (more about that later) or go to a faster EPROM.

Calculating the timing for other flash memories is the same as for EPROMs, except that you also must take into account the write timing.

RAM

Two general types of RAM are used in embedded systems. The first and most common is *static* RAM (SRAM). Static means that the memory cells do not change unless they are rewritten or the power is removed. A static RAM consists of an array of flip-flops that are selected by a decoding array inside the chip. Static RAM usually comes in x8 configurations, but there are some x16 devices.

A special case of static RAM is *nonvolatile* RAM (NVRAM). This consists of a special low-power RAM chip packaged with a battery (usually lithium). The combination also includes power-switching circuitry that operates the RAM from system power when available and from the battery when system power is removed. The switching logic also protects the RAM from inadvertent writes when the power is below a certain threshold, usually when the system power is coming on or going off.

The other type of RAM is *dynamic* RAM (DRAM). Dynamic RAM is used in personal computers (PCs). It stores information as charge on a tiny capacitor, one per data bit. Because the capacitor charge bleeds off, the data must be refreshed periodically. DRAM multiplexes the address pins into row and column addresses. The row address is latched in with a signal called *RAS* (row address strobe), and the column address is latched in with a signal called *CAS* (column address strobe). The need to multiplex the addresses, generate the strobes, and refresh the part make DRAM more difficult to design with. Dynamic RAM can be made smaller than static RAM, so a single DRAM chip will be denser than a corresponding static RAM chip.

Calculating RAM Access Time

Figure 2.8 shows an SRAM IC connected to a microprocessor with a multiplexed address/data bus. Note that the connections are identical to those for an EPROM with the exception of the added write enable signal, which is connected to the microprocessor -WR signal. Although not shown, some RAM ICs have multiple chip select inputs.

For static RAM, and during a read cycle, RAM timing is calculated the same as EPROM timing. For a write cycle, additional factors must be considered.

First, the data and control setup and hold times must be calculated. Figure 2.9 shows a static RAM write cycle. Several additional timing parameters have to be taken into consideration with a RAM:

Address setup time. Unlike an EPROM, the contents of a RAM can be changed. The RAM requires that the address be stable before the write strobe (-WR) is asserted. If the minimum setup time is not met, the address decoding logic inside the RAM still may be changing when -WR

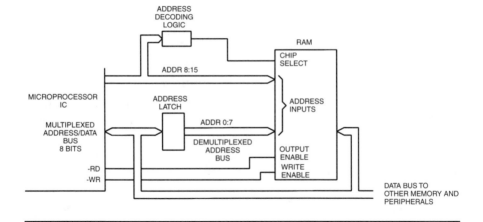

Figure 2.8
RAM Connected to a Microprocessor with a Multiplexed Address/Data Bus.

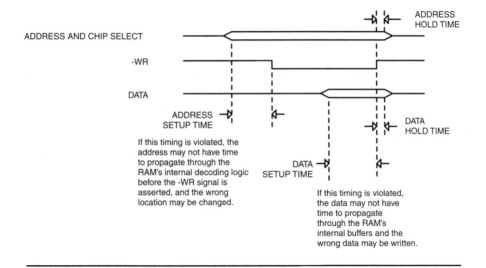

Figure 2.9
Static RAM Write Cycle Timing.

is asserted, and the wrong address or multiple addresses may be changed. Note that the address setup time applies to the *leading* edge of the -WR strobe.

Data setup time. To guarantee that the correct data are written to the selected location, the data must be stable before the trailing edge of the -WR signal. This provides time for the data to get through the RAM's internal delays.

Data/address hold times. The data and address must each be held for some specific time after the trailing edge of -WR. This guarantees that the negation of the -WR signal has time to propagate through the RAM's internal delays before the address and data change.

The price for not meeting these parameters is intermittent RAM problems—locations that seem to change at random or data that are incorrectly written. Like EPROM access time problems, the symptoms may occur only with specific brands of parts or only when the temperature reaches a certain point.

Calculating the address setup time is as follows: Using the microprocessor data sheets, calculate the time that the address is stable prior to assertion of the -WR signal (remember: *leading* edge). Subtract address latch propagation delays. The result must be greater than the address setup time specified for the RAM chip to be used. If it is not, you must either select a faster RAM or delay the assertion of -WR using external logic. The formula for this is

$$Tasr = Tasm - Td$$

Tasr is the address setup time required for the RAM; Tasm is the address setup time provided by the microprocessor; Td is any delays in the data path, such as a data bus buffer.

Note that delays in the -WR path do not affect address setup time. In fact, a delay in the -WR path *improves* address setup time since it gives the address more time to stabilize at the RAM before the -WR signal arrives. However, this is not a free lunch—delays in the -WR signal path can cause a data hold time problem, which we look at later.

Data setup time is calculated in much the same way as address setup time. Calculate the time from when microprocessor asserts the data until the *trailing* edge of the -WR signal. Subtract any data bus buffer delays. Your RAM must have a data setup time that is *less* than the calculated value:

$$Tdsr = Tdsm - Td$$

where Tdsr is the data setup time required by your RAM; Tdsm is the data setup time, prior to the trailing edge of –WR, provided by the microprocessor; Td represents any delays in the data path, such as a data bus buffer.

Data and address hold time are calculated by determining how long the microprocessor holds the address and data after the trailing edge of -WR. If you use address latches for all address lines, address hold time usually will not be a problem, since the address will remain stable until the start of the next cycle. If you have data bus buffers, *add* the *minimum* propagation delay, if known, to the microprocessor data hold time. If the minimum is not known, do not add the buffer delay. If there are delays in the -WR path, subtract those, as they delay removal of -WR from the RAM. The RAM must have a smaller hold time requirement than the calculated result:

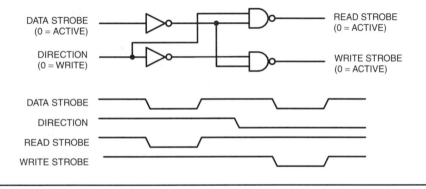

Figure 2.10
Generating Independent Read and Write Strobes from a Microprocessor That
Produces Data Strobe and Direction Signals.

$$Tholdr = Tholdm + Td$$

where Tholdr is the data hold time required for RAM; Tholdm is the
data hold time provided by microprocessor; Td is the minimum data bus
propagation delay (if known) plus delays in the -WR path (if any).

The preceding information is based on the assumption that your microprocessor generates separate -RD and -WR signals. For microprocessors, such
as the Z8 family, that generate a data strobe and a R/W signal, there are two
options: First, the -OE pin on the RAM is grounded and the -WE signal is connected to R/W. One of the chip select signals is connected to the data strobe
from the processor. The -WE signal on a static RAM overrides the -OE signal,
permitting a write cycle to occur even if the -OE signal is low. The disadvantage to this is that the output enable time becomes the chip select access time,
which may require that a faster device be used.

The second option for these processors is to generate the read and write
strobes from the microprocessor data strobe and direction signals. Figure 2.10
shows a typical circuit for doing this.

Nonvolatile RAM

As mentioned earlier, NVRAM usually is an SRAM with a battery and
power switching logic added. It has the same timing parameters as SRAM and
is interfaced in the same way.

Dynamic RAM

Dynamic RAM, as mentioned earlier, stores information as a charge on a capacitor. DRAM is less common in embedded designs than static RAM and typically

is used where a lot of memory is needed. Because a DRAM memory cell consists of a capacitor and a transistor, while an SRAM cell requires a flip-flop, DRAM density for a given level of technology will be higher than SRAM. At this time, 128 K × 8 (1 MB) SRAMs are common, while common DRAMs are 4 MB × 4. The disadvantages of DRAMs are, first, that interfacing is more difficult and, second, that the parts have to be refreshed periodically.

A typical DRAM has half as many address lines as are needed to access the entire memory array. The lines are multiplexed with the row address presented first and then the column address presented on the same pins. A 4 MB RAM has 4,194,304 locations and requires 22 address inputs. The actual DRAM would have 11 address lines.

DRAM timing is less forgiving than SRAM timing. A DRAM has several important parameters:

Row address setup time. The time that the row address must be stable on the address inputs before -RAS is driven low.

Row address hold time. The time that the row address must be stable after the falling edge of -RAS.

Column address setup time. The time that the column address must be stable on the address inputs before -CAS is driven low.

Column address hold time. The time that the column address must be stable after the falling edge of -CAS.

RAS access time. The maximum time from the falling edge of -RAS to output data available.

CAS access time. The maximum time from the falling edge of -CAS to output data available.

RAS hold time. The minimum time that -RAS must remain low after the falling edge of -CAS.

RAS/CAS precharge time. The times that -RAS and -CAS must remain high before the next cycle can start.

Looking at a DRAM data sheet reveals many more timing parameters than those listed here, but these are the key ones. Note that two access times are listed: RAS and CAS. The actual access time is determined by the circuit. In a fast circuit, CAS may enable the output buffer before the logic in the DRAM has decoded the row address, and the RAS time becomes the actual access time. In a slower circuit, where the row address will be internally decoded by the time CAS occurs, the access time will be governed by when CAS falls. To put it another way, data will not be available from the DRAM any sooner than the RAS time after the falling edge of RAS, even if the address multiplexing and CAS timing are very fast.

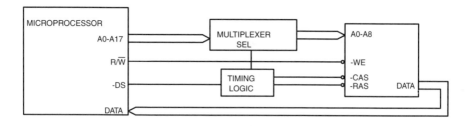

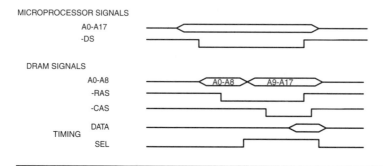

Figure 2.11
DRAM Interface to a Motorola-Type Bus.

Figure 2.11 shows a hypothetical $256\,\mathrm{K} \times 8$ DRAM connected to a microprocessor with a Motorola-type bus. Actual DRAMs are usually $\times 1$ or $\times 4$; this example is $\times 8$ for simplicity. Also, the data transfer acknowledge (DTACK) timing does not appear on this figure for the same reason.

The address is presented to the DRAM through a multiplexer. At the start of the cycle (see Figure 2.11), the low-order address bits (A0–A8) are passed through to the DRAM and -RAS is pulsed, latching the row address into the DRAM. After the address hold time is met, the SEL line to the multiplexer is toggled, causing the high-order address (A9–A17) to be presented to the DRAM. After the column address setup time is met, -CAS is pulsed, latching the column address. Data from the DRAM is available after the CAS access time.

The direction signal (R/W) is passed directly to the DRAM. If the WE pin on the DRAM is low when -CAS goes low, the DRAM will start a write cycle. If WE goes low *after* -CAS goes low, the DRAM will do a read cycle, driving read data onto the data bus, followed by a write cycle. This is called a *read modify write* (rmw) *cycle*. Write data is latched on the leading edge of -WE or -CAS, whichever is later. Few embedded processors execute rmw cycles. The reason this timing is important is to avoid bus contention for processors where the write signal may be later than -CAS. Note, however,

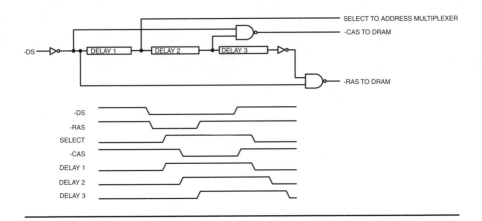

Figure 2.12
Typical DRAM Timing Logic.

that the data is latched and must be stable before -WE or -CAS, whichever occurs later.

Figure 2.12 shows a method of implementing the timing logic for Figure 2.11. The address setup time from the processor, prior to the leading edge of -DS, meets the DRAM row address setup time, so -RAS can go active with -DS. After Delay 1, which is the row address hold time, the select signal to the multiplexer changes states, which switches the DRAM address inputs from the row to the column address. After Delay 2, which is the column address setup time, -CAS is driven low. -RAS goes back high after Delay 3, which is the row address hold time. -RAS could be held active throughout the entire memory cycle, but removing -RAS after -CAS is asserted makes it easier to meet the -RAS precharge time.

When -DS goes inactive, -CAS is removed immediately. The DRAM drives the data bus as long as -CAS is active, allowing the -CAS inactive state to propagate through the delays could cause bus contention.

The delays in Figure 2.12 may be implemented with delay lines or synchronous logic. In either case, you must make sure that the inactive state of -DS has propagated through all delays before the next cycle starts. This circuit is simplified, since it does not include a provision for separate refresh, but it shows the timing principles involved.

Because DRAM has two address setup/hold times and two address strobes in one cycle, it is slower than equivalent SRAM parts. The example in Figure 2.11 did not start the DRAM cycle until the data strobe from the processor occurred. This may require the addition of wait states, depending on processor and DRAM speed. It is possible to start the cycle early. On Intel-type processors, the -RAS signal can be generated when ALE goes active. With a

Embedded Microprocessor Systems

Motorola-type bus, the address strobe can be used to start the cycle. In both cases, the address decoding must be fast enough to ensure that the RAM is not falsely selected. Also, the address multiplexer adds an additional level of delay that must be taken into account; the row address must be stable prior to the leading edge of -RAS.

Dynamic RAM must be refreshed. The storage capacitor loses its charge fairly quickly, typically in 15 milliseconds (ms) or less. Refresh is accomplished by accessing each row in the DRAM. Internal logic in the DRAM restores the charge on the capacitor. Note that accessing any row refreshes all columns in that row. For example, a 256K DRAM typically has 256 rows and 1024 columns. Any read or write cycle refreshes the entire row, but the catch is that *all* rows (that is, all row addresses) must be refreshed within the refresh interval.

Unless refresh was accomplished with an actual data read, early DRAMs required that the user generate a refresh address and a -RAS signal every 15 microseconds (µs) or so. On a 256K DRAM, this refreshes all 256 rows in about 4 ms. This scheme required an external counter and a way to multiplex the count onto the address lines. The timing logic had to recognize a refresh request and generate a refresh cycle, arbitrating it with processor cycles. Newer DRAMs can still use this -RAS-only refresh, but they make refresh easier by also supplying an internal refresh address counter. Each time the DRAM is refreshed using a special refresh cycle, the counter increments to the next address.

The internal refresh cycle is started by reversing the order of -CAS and -RAS. -CAS is driven low first, followed by -RAS. The DRAM recognizes this condition and refreshes the internal row, then increments the refresh counter. The data bus is not driven during the refresh cycle.

While an external counter is not required for the internal refresh cycle, refresh still poses some problems. First, an external timer must generate a request for refresh at regular intervals. Second, the interface logic must interleave the refresh cycles with the processor access cycles. What happens if the DRAM is in the middle of refreshing and the processor wants to start a read cycle? There are several ways to handle the conflict between processor and refresh cycles:

Use wait states. If the processor wants to use the DRAM, it must wait until the current refresh cycle is completed. This probably is the most common method of handling refresh.

Synchronize refresh to the processor. Allow refresh to occur only for cycles that do not use the DRAM. This can be dangerous if the processor is executing code from the DRAM, which may never permit refresh to occur. However, if the DRAM will be used only for data, this approach may be feasible. A slow processor may permit the entire refresh cycle to be performed without affecting normal operation, such as during the ALE time.

Use the direct memory access (DMA) capability of the processor. DMA can be used for refresh by allowing the refresh logic to request a hold and do the refresh cycle when the processor acknowledges the hold request. The disadvantage of this is that it usually takes a few clocks for the processor to get in and out of hold.

Use built-in refresh. Many microprocessors, such as some versions of the 80C186, have built-in refresh logic. This consists of an internal timer that generates refresh requests at regular intervals. Processors that generate refresh requests internally also provide the refresh row address, so that -RAS-only refresh cycles may be performed.

If the internal refresh capability of the DRAM is to be used, the DRAM timing logic must detect the refresh condition and generate the CAS-before-RAS cycle. DRAM timing logic may be implemented using discrete logic or programmable logic devices (PLDs). The required delays may be generated using delay lines or a clock. Either way, *all* the DRAM timing constraints must be met. Probably the most common mistakes in DRAM design are failing to meet the setup/hold times and failing to meet the precharge times, especially when switching between refresh and processor access to the part.

If you do not want to roll your own timing logic, a number of controller ICs simplify the task of interfacing to and controlling DRAMs. Typical examples are the DP8440 and DP8441 from National Semiconductor. But the decision to use DRAM implies a considerable increase in the cost and complexity of a design and should be considered carefully to see if it is really necessary.

This has been a lengthy discussion of connecting memory to a microprocessor and calculating the worst-case timings, but it is important because the timing of all other peripherals is calculated in the same way. The foregoing information is based on the assumption that the designer will use worst-case numbers. Some manufacturers provide a table or other information that indicates the memory speed needed for a specific clock rate. But, if it is not specified that way, assume the worst possible timing scenario eventually will happen.

One last note about timing calculations: They are straightforward to do with a calculator, but a number of timing analysis programs for PCs will do the calculations, display the resulting waveform on the screen, and even highlight problem areas in red. Examples of these are *Timing Diagrammer Pro* from Synapticad and *Tau* from Mentor Graphics. These programs typically include libraries of microprocessors and other parts, including the timing parameters, so you need not even look up the worst-case parameters on the data sheets. The program does all the calculations for you and you can print out a timing diagram that can be included in the board specifications or other documentation.

I/O

The entire point of an embedded microprocessor is to monitor or control some real world event. To do this, the microprocessor must have I/O capability. Like a desktop computer without a monitor, printer, or keyboard, an embedded microprocessor without I/O is just a paperweight. The I/O from an embedded control system falls into two broad categories: digital and analog. But at the microprocessor level, all I/O is digital. (Some microprocessor ICs have built in A/D converters, but the processor itself still works with digital values.)

The simplest form of I/O is a register that the microprocessor can write to or a buffer that it can read. Figure 2.13 illustrates these two implementations. When the microprocessor performs a read to the address of the 74AC244, the decoding logic produces a read strobe and the 74AC244 outputs are enabled

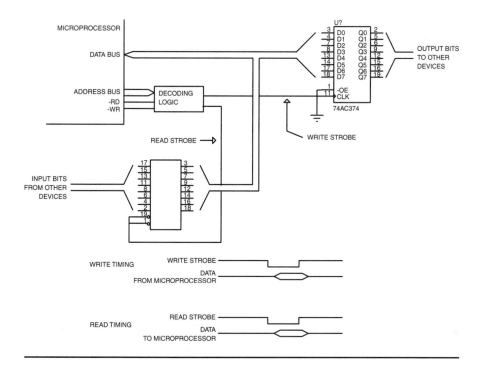

Figure 2.13
Simple Input and Output Ports.

onto the microprocessor data bus. Similarly, a write to the address of the 74AC374 generates a write strobe that clocks the data bus value into the 74AC374. The input bits to the 74AC244 could be switch contacts, a temperature sensor, comparator outputs, or any other digital information. Similarly, the 74AC374 outputs could drive LEDs, or a relay, or other logic.

The decoding logic to generate the strobe signal can be implemented with PLDs, discrete logic, or demultiplexers such as the 74AC138/139. The decoding logic should produce output strobes that follow the microprocessor -RD and -WR signals.

Figure 2.14 illustrates three decoding circuits. The first, an 8-input NAND gate, decodes address lines A8–A15, the upper eight lines on a 16-bit address bus. When the microprocessor accesses any location in the (hex) range FF00 to FFFF, A8–A15 will all be high, producing a low at the NAND gate output. Of course, a wider address bus or a need to decode to greater resolution will require a wider NAND circuit.

The second circuit in Figure 2.14 is a 74AC138. This circuit produces output strobes that follow the data strobe from the microprocessor and are

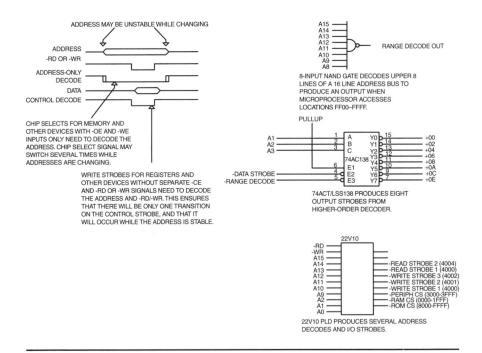

Figure 2.14
Address Decoding Circuits.

suitable for clocking data into a register or for enabling a buffer. The select inputs (A, B, and C) are connected to the microprocessor address bus, bits A1–A2. One enable is connected to a range decode (such as the NAND gate later in Figure 2.15), and a second enable is connected to the microprocessor data strobe. The unused enable is pulled up. As indicated in Figure 2.14, the eight outputs of this circuit each go active at a different offset from the start of the range decode. For example, if the range decode was active from addresses FFF0 to FFF7, the outputs of the 74AC138 would go active at addresses FFF0, FFF2, FFF4, and so forth.

One drawback to this circuit is that each strobe goes active for either a read or a write. To get independent read/write strobes, two 74AC138 circuits are used. Instead of the data strobe input, one 74AC138 is enabled with -RD and the other with -WR.

The reason it is necessary to gate the I/O strobes with the -RD or -WR signals is because the address typically is held longer than the data for a write. If a write strobe was just an address decode (not gated with -WR), the register would not get a clock until the after data were gone. If the read strobe were not gated with -RD, an output buffer would be enabled too long and there may be bus contention at the end of the bus cycle when the next one starts and the microprocessor tries to drive the data bus. A second reason for gating the strobes is that, while the address is changing at the start of a bus cycle, the address lines may skew and not all change at the same time. Consequently, the wrong address may momentarily appear on the address lines, and the wrong device could be selected. The decoding logic could produce a short pulse on a write strobe signal, clocking garbage data into a register. Gating read and write strobes with the control signals makes sure the strobes go active only when address and data signals are stable. Figure 2.14 shows this timing.

The last circuit in Figure 2.14 shows how a 22V10 (or other PLD) can be used to generate address decodes and read/write strobes from a single IC. This example decodes a 16-bit (64 K) address space, producing a 32 K EPROM chip select from addresses 8000–FFFF, an 8 K RAM chip select from 0000–1FFF, and a peripheral chip select from 3000–3FFF. Read strobes are generated at 4000 and 4004, and write strobes are generated at 4000, 4001, and 4002.

Since the EPROM, RAM, and our hypothetical peripheral IC have their own -WE and -OE inputs, the chip selects for these parts will not be gated with the -RD and -WR signals from the microprocessor. The read/write strobes will be gated with the control signals, however, because they are intended for clocking data into a latch or for enabling a buffer.

The following equations implement this PLD in CUPL/ABEL format (& is the logical AND function, # is the logical OR function, a ! prefix indicates a low-true signal, a double slash [//] precedes comments).

```
!EPROMCS = A15;    // 8000–FFFF
!RAMCS = !A15 & !A14 & !A13;    // 0000–1FFF
!PERIPHCS = !A15 & !A14 & A13 & A12;  // 3000–3FFF
!WSTB1 = !A15 & A14 & !A13 & !A12 & !A11 & !A10 &
!A9 & !A2 & !A1 & !A0 & !WR;    // 4000
!WSTB2 = !A15 & A14 & !A13 & !A12 & !A11 & !A10 &
!A9 & !A2 & !A1 & A0 & !WR;    // 4001
!WSTB3 = !A15 & A14 & !A13 & !A12 & !A11 & !A10 &
!A9 & !A2 & A1 & !A0 & !WR;    // 4002
!RSTB1 = !A15 & A14 & !A13 & !A12 & !A11 & !A10 &
!A9 & !A2 & !A1 & !A0 & !RD;    // 4000
!RSTB2 = !A15 & A14 & !A13 & !A12 & !A11 & !A10 &
!A9 & A2 & !A1 & !A0 & !RD;    // 4004
```

LSI I/O

The advantage of using discrete latches and buffers for I/O is simplicity. The disadvantages are:

Unidirectional operation. The latch outputs cannot be read to determine if a particular bit is set.

Not programmable. The inputs are always inputs, the outputs are always outputs. If you need nine inputs instead of eight, but only seven outputs, you cannot use a latch output as an input—you have to add another 74AC244 buffer.

PC board real estate. Each new set of eight inputs or outputs requires another IC and another output from the decoding logic.

Interface. The requirement for discrete read/write strobes to each device complicates interface with 68000- or Z8-type processors that generate a common data strobe and direction signal.

In addition to these, another problem is that this type of discrete I/O is limited to just that—digital I/O bits. A design often requires other functions, such as a timer, serial interface, or analog-to-digital (A/D) converter, which cannot be implemented with simple latches.

Peripheral ICs

Most microprocessors intended for multichip designs have peripheral ICs as part of the product family. These include timer/counters, serial interface chips, and port expansion. A few examples are described here.

Timers

A timer peripheral consists of a counter that decrements or increments at some clock rate. The processor can read the count and the timer may generate an interrupt or pulse an output pin when the count rolls over to 0. Some timer ICs allow one timer to be cascaded from another for long delays. The timer output varies with the particular IC used; many have outputs that can be programmed for a square wave, single pulse on output, or variable duty cycle. In addition to the count, the processor can control timer start/stop and modes of operation. Typical uses for a timer IC are to generate a delay, usually for scheduling some real-time event, controlling motors (DC PWM or stepper), and generating a regular timekeeping interrupt.

I/O Ports

These ICs provide a multichip design with the same programmable I/O port capability as a microcontroller. A typical I/O port IC may provide three or four 8-bit ports. Some port ICs include hardware handshaking that permits a port to be used for interprocessor communication in multiprocessor systems. The processor can control the direction of each port (sometimes to the bit level, depending on the part), and all modes of operation. I/O port ICs are also called *port expanders.*

Interface ICs

These provide standard interfaces, such as SCSI, IEEE-488, asynchronous serial I/O, Ethernet, or "Firewire." Many of these parts handle more than one interface. Some UARTs (universal asynchronous receiver/transmitters), for example, can handle multiple serial protocols, relieving the processor from the burden of handling each received byte.

Interrupt Controllers

Interrupt controllers simplify adding interrupts to processors. See Chapter 4.

IC Functions

In many cases, an IC may combine two or more functions. Table 2.1 is a brief list of typical LSI (Large Scale Integration) I/O parts.

Most parts that are designed for a specific family of processors can interface fairly painlessly to parts in that family. However, sometimes the designer needs a function that is performed by a peripheral part from another family. For example, you might want an interface between a Motorola peripheral and an Intel processor. Although showing every possible combination of peripherals and microprocessors would take volumes, some representative examples are illustrated here.

Figure 2.15 shows a simplified timing diagram for the Intel 80188 and the Motorola 68230. On a read cycle (the address multiplexing does not appear on the diagram for simplicity), the 80188 generates a -RD strobe and the peripheral is expected to have data available at the trailing edge of -RD (actually at the clock edge that precedes the end of -RD). For a write cycle, the 80188 produces a -WR strobe and the write data is stable some time before the trailing edge of -WR.

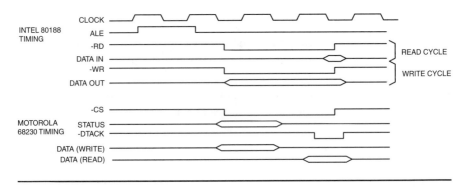

Figure 2.15
Intel 80188 versus Motorola 68230 Timing.

Table 2.1
Typical LSI I/O ICs.

Part	Function	Family
Intel 82C54	Three timers	All Intel processors
Intel 8259	Interrupt controller	Intel processors
Intel 8255	Four I/O ports	Intel processors
Zilog Z84C20	Two 8-bit ports	Zilog processors
Zilog Z8530	Serial communications	Non-Zilog processors
Zilog Z8536	Three I/O ports, timers	Non-Zilog processors
Motorola 68230	Parallel I/O ports	68000 family
SGS 68564	Dual UART	68000 family
National LM628	DC motor controller	Intel processors

Embedded Microprocessor Systems

The 68230 has five register select inputs that are used to address internal registers. These are connected to the microprocessor address lines. The 68230 expects a -CS signal, which is a logical AND of the address decode and data strobe. The register select (address) inputs, the R/W signal, and the write data (for a write cycle) all must be stable prior to the leading (falling) edge of -CS. The 68020 also expects the processor to hold -CS active until it returns -DTACK.

While the 80188 has stable address and read/write status at the leading edge of the -RD or -WR signals, the write data is not stable at the leading edge of -WR. Also, the 80188 does not use a -DTACK signal to terminate the bus cycle.

Connecting a 68230 to an 80188 requires the following changes:

The -RD and -WR signals must be converted to a single -DS data strobe.

For a write cycle, the synthesized -DS signal must be delayed until the data on the bus is stable.

The 80188 must be forced to hold the bus signals active until the 68230 returns -DTACK.

Figure 2.16 shows how this conversion can be performed. One of the internal 80188 chip selects (PCS0–PCS5) generates the range decode and 80188 address lines A0–A4 drive 68230 RS1–RS4. The -RD and -WR signals from the 80188 are gated together to produce a single -DS to the 68230. The -WR signal is delayed by one half-clock cycle to ensure stable data at the leading edge of

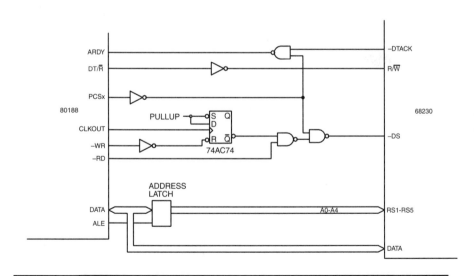

Figure 2.16
Connecting a 68230 to an 80188.

-DS. The 80188 DT/R signal is inverted to drive the 68230 R/W input. Finally, the -DTACK signal is returned as a -WAIT signal to the 80188 ARDY input. ARDY is driven low (not ready) as long as the 68230 is selected and -DTACK is false.

Z80 Peripherals

The Z8536 and Z8530 peripherals are popular parts. Interfacing these to an Intel processor is simpler than interfacing to 68000 family peripherals. The Z853x parts have separate -RD and -WR inputs but, like the 68000 family parts, the write data must be stable at the leading edge of -WR.

Figure 2.17 shows the circuitry necessary to connect a Z8530 or Z8536 to an 80188 or other Intel processor. Like the 68230 interface, the Z853x interface uses a PCS (Peripheral Chip Select) line for the range decode and delays the -WR signal until write data are stable. One important addition is the AND gates between the -RD and -WR signals from the 80188 and the corresponding inputs to the Z853x. This is added because the Z853x parts interpret assertion of both -RD and -WR as a reset condition. The AND gates drive both inputs to the Z853x low when the RESET output from the 80188 goes active.

The circuitry in Figure 2.17 could be used with a PIC17C4x processor as well, except that the -WR delay is not required. The -WR signal connects directly to the -WR AND gate, like the -RD. Interfacing an Intel peripheral to a Motorola processor is probably a less common requirement, but Figure 2.18 shows how.

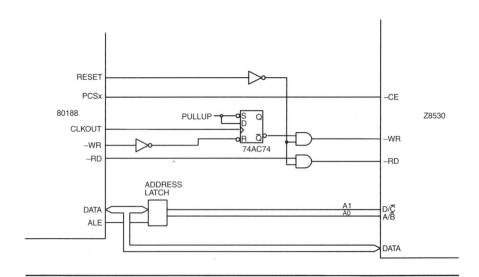

Figure 2.17
Connecting a Z853x to an 80188.

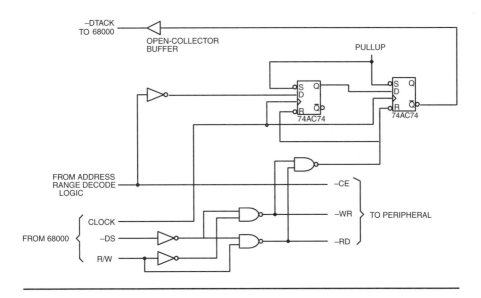

Figure 2.18
Connecting an Intel Peripheral to a Motorola Processor.

The -DS signal from the Motorola processor is split into separate -RD and -WR signals using the R/W signal. A data strobe with R/W high produces a -RD to the peripheral and a data strobe with R/W low produces a -WR. The 68000 family parts require a -DTACK to terminate the cycle; this is generated with a pair of 74AC74 flip-flops. The -DTACK is returned when -RD or -WR occurs with -CE, indicating an access to the peripheral. Two 74AC74 gates are shown; the actual delay required to guarantee proper operation depends on the relative speeds of the processor and peripheral. Of course, like any of the examples in this chapter, this circuitry could be embedded in a PLD.

Data Setup/Hold Time Problems

Wait states, which were covered earlier, will extend a read or write cycle enough to allow a fast microprocessor (such as a 20 MHz 80C186) to interface with a slower peripheral (such as a 6 MHz Z8530). But wait states alone will not solve all mismatch problems. Some peripherals require longer setup or hold times for address or data than the processor provides. Addition of wait states does not affect these timings, because wait states extend only the processor cycle without affecting the timing of asserting and removing the control strobes. If there is a mismatch in this area, additional logic must ensure that all parameters required by the peripheral and the processor are met. This can be determined only by examining the published timing information for the processor and the peripheral. In most cases, setup and hold time problems can be fixed by a combination of wait states and manipulation of -RD and -WR or -DS before the signals reach the peripheral, either by delaying them or terminating them early. The following example will illustrate this.

Extended Data Hold Time

Occasionally you run into a peripheral IC that requires write data to be held for some time after the WRITE strobe goes away. A typical example is the LM628 motor controller IC from National Semiconductor. Most microprocessors do not guarantee a data hold time long enough for parts that need an extended data hold time. There are two ways to implement an interface to parts like this. The first method is to latch the data and leave the buffer enabled to the peripheral device (Figure 2.19). This provides the fastest transfer speed, since the processor timing is unaffected. However, it requires that you implement a bidirectional, latching data buffer for the peripheral part.

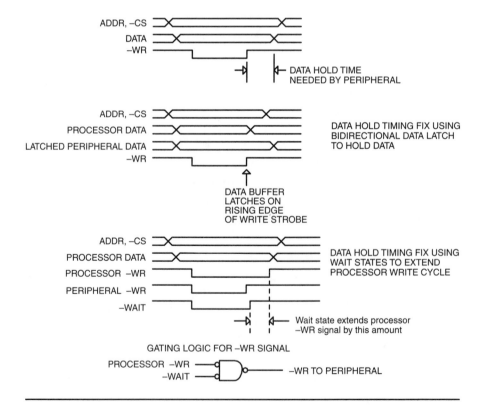

Figure 2.19
Extended Data Hold Time.

Also, the data latch works only if the processor cannot do another write quickly enough to change the latch contents too soon and violate the timing requirements. You might see this if the peripheral with extended hold requirements is memory-mapped and the processor performs a word write as a pair of back-to-back byte writes.

The second, and simpler, method for extending the data hold time also is shown in Figure 2.19. Here, a wait state is used to extend the processor write cycle. The -WR pulse to the peripheral device does not connect directly to the processor -WR signal but instead goes through some intermediate logic. This logic terminates the -WR signal to the peripheral early in the cycle, when the wait request is removed. Since the processor will extend the cycle one clock past this point, the data will be held on the bus for the peripheral device. This ensures that the data hold time requirement is met. It also guarantees that the processor will not perform a second write that violates the chip timing.

8- versus 16-Bit Interface

Some processors are available with both 8- and 16-bit external interfaces. For example, the 80188 uses the same microprocessor core as the 80186. But the external interface to memory and I/O is only 8 bits wide on the 80188, versus 16 bits on the 80186. Similarly, the 68008 microprocessor has a 16-bit 68000 CPU core, but interfaces to external memory and I/O devices via an 8-bit bus. The Siemens/Infineon C167 can be programmed for either 8- or 16-bit external memory operations.

The drawback to using an 8-bit bus is performance. While the internal CPU is the same as the 16-bit sibling, external access is slower. A processor with an 8-bit external interface requires two memory cycles to get a 16-bit word, where the 16-bit bus can get a word in one cycle.

So why would anyone want to use a processor with an 8-bit bus? Cost. Using a 16-bit bus requires two of everything. EPROMs and RAM must be 16 bits wide. Your program may require only 1 K of code space and 256 bytes of RAM, but you still need a 16-bit interface to the processor, which means two (or more) RAM and EPROM chips.

Some RAM and ROM ICs feature 16-bit data buses, but they typically are more expensive than their 8-bit counterparts. In addition, many peripheral ICs only have an 8-bit bus, which gives the 16-bit processor less of an edge. On the other hand, some peripherals require 16-bit interfaces, which precludes use of an 8-bit processor. An 8-bit processor *could* be used but only with data latches and external logic to turn two 8-bit cycles into a single 16-bit cycle at the peripheral.

The 16/8-bit concept also applies to other bus widths. The Intel 80C960SA is a 16-bit, multiplexed-bus version of the 32-bit 80C960. The 386EX is a

16-bit bus version of the 386 processor, optimized for embedded applications. Both the 80C960 and the 386EX use the same processor core as the larger parts they are derived from; only the external bus is narrower.

16-Bit Considerations

The interface examples shown so far have been for 8-bit processors. Interfacing to 16-bit processors is similar, except for the wider bus width. However, some 16-bit and wider processors require somewhat more complex interfacing, since they can execute both 8- and 16-bit cycles.

The 80186 has a 16-bit bus, but if a word-wide (16-bit) memory access is performed to an *odd* address, the processor will perform two back-to-back 8-bit cycles to access the word. This is important because the processor expects to operate on only 8 bits at a time; the remaining 8 bits are unused. Say that memory location 0006 contains the value 1B2C. The CPU may access this as a 16-bit value or it may access either the high (1B) or low (2C) bytes. The CPU can write the lower byte to 3D, leaving the result 1B3D.

Two signals (A0 and BHE) on the 80186 control which byte of memory or I/O is accessed (low, high, or both). If the odd-address example happens to be a write and the memory design assumes that all accesses will be 16 bits wide, then each of the two 8-bit writes will write invalid data to 1 byte of the memory word. The memory logic must decode the BHE and A0 signals to determine if just 1 byte is being written. In the example just given, if the logic does not properly decode the A0 and BHE signals, then writing 3D to the least significant byte of the word will result in xx3D instead of 1B3D. The xx is an unknown value—the most significant byte will be whatever data is on the bus when the write occurs.

Other processors have similar characteristics. The 386EX is a 32-bit processor with a 16-bit external bus, but it also can perform 8-bit cycles. When designing with a processor that has 16 (or more) bits and can perform byte-oriented cycles, be sure the memory design handles these operations correctly. If you are using 2-byte-wide ICs to implement a 16-bit-wide memory, you can gate the write signal to each memory IC with the appropriate byte select signal. On the 80186, BHE would be used to gate the most significant byte, and A0 would be used to gate the least significant byte. Figure 2.20 shows the gating needed to control the -RD and -WR lines when two SRAMs are connected to an 80186 processor.

In the figure, the chip selects for both RAM ICs come from the 80186 -LCS output, and both chip selects are connected together. The logic enables the low RAM IC when A0 is low, and the high RAM IC when BHE is low. When both signals are low, both devices are enabled. Note that the address inputs to the RAM ICs get address lines A1–A15. This is a typical usage of the address

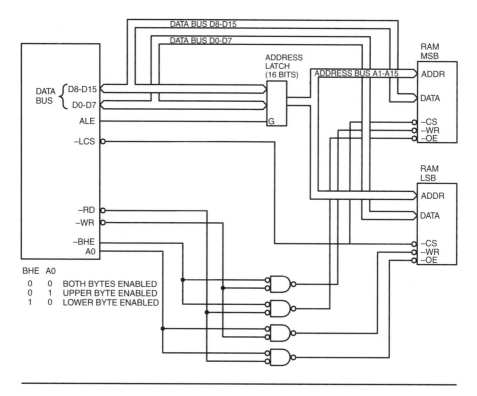

Figure 2.20
Bus High/Low Gating Logic.

lines on processors that have 16-bit or wider data buses. When interfacing to the 8-bit bus of the 80188, the RAM and ROM ICs will get all the address lines, including A0. On the 16-bit 80186, A0 is used as a byte selector. Note that both processors can access the same amount of memory (1 MB), but the 186 can access 16 bits at a time.

Microprocessors that can access memory in cycles that have fewer bits than the bus width (16-bit CPU doing 8-bit cycles, 32-bit CPU doing 16-bit cycles) will have status bits that tell the memory decoding logic what size the access is. A 16-bit CPU that can perform 8-bit accesses needs 2 bits (high byte, low byte, both bytes) to determine what portion of the bus is used. As mentioned, on the ×86, these are BHE and A0.

A 32-bit processor that can access 8, 16, or 32 bits would typically have seven possible states (byte 0, 1, 2, or 3, low word or high word, 32 bit access) so three status signals would be needed. Theoretically, there are other combinations, such as a 16-bit word composed of bytes 0 and 2, but these normally are not available. In any event, the decoding logic must decode the status lines

Hardware Design

65

and route the right data to the data bus. Just like writing a single byte of a 16-bit word, this is especially important when writing to memory, as the CPU may want to write to only 1 byte of a 32-bit-wide memory device. It is important not to corrupt the other 3 bytes during that write cycle.

In many cases, a 16-, 32-, or 64-bit processor will interface to an 8-bit device. An example of this is the 16550 UART, which is common in the PC world. To simplify the design, the hardware normally is designed to support only 8-bit access to the device, using 8 bits of the data bus. If the software attempts to perform a word write or a byte write to the wrong address, the result is undefined. Interfaces like this normally do not attempt to cover all the cases, since the software can be written to avoid invalid accesses.

Data Bus Loading

A microprocessor is specified to drive a particular DC loading (sourcing or sinking current) and a particular capacitance loading. A common mistake is to ignore these parameters and assume that the processor will drive the bus. This is a dangerous practice, especially if a failure is likely to result in the engineer having to fix it in an unsavory place, like an oil field or a country where you should not drink the water.

Bus loading problems can result in much the same sort of symptoms as setup/hold time violations. In fact, bus loading problems can *cause* setup and hold time problems, because they change processor timing.

A microprocessor is specified to meet its performance characteristics with maximum DC sink and source currents and with a maximum load capacitance. AMD's version of the 80C188, for example, specifies a sink current of 2 mA and a capacitance drive capability of 100 pF. If you exceed these numbers, the performance of the part starts to degrade.

When the standard interface logic was LSTTL or FTTL, DC loading usually was the primary problem. Now that the world has shifted primarily to CMOS, I see more problems with capacitance. I think designers look at the extremely low leakage of CMOS inputs and just forget that those inputs have capacitance.

Some parts, such as the 80C188, have a derating chart for capacitance, which shows how much the outputs are slowed by added capacitance beyond the specified value. However, in all cases, whether specified or not, excessive loading can cause problems.

To calculate DC loading, add the maximum sink and source currents required by all inputs and compare them to all the outputs (including bidirectional devices). The sum of the input currents must not exceed the capability of the device with the smallest output drive capability. On CMOS devices, check not only the output current capability but what sink current does to the output voltage. The output current of some CMOS devices is specified at TTL level voltages. If one of those devices is driving an IC that requires CMOS-level input voltages, there may be a problem. If the total DC loading pulls the output of the first device down, the second part may not see the correct value.

Capacitance loading is similar. Add up the input capacitance (sometimes specified as I/O capacitance for bidirectional devices) and compare it to the drive capability for each device that has to drive the bus. The total capacitance should be less than what the device with the lowest drive specification can tolerate. If derating curves are provided, they can be used to determine if access times are degraded enough to be a problem.

If a loading problem is discovered, the simplest fix is to add a buffer to the data or address buses. This isolates the processor bus from the peripherals that load it. The problem with a buffer is that it adds delay to the system. If timing is marginal, a faster EPROM, for example, may be required.

Note that adding a buffer may just move the problem around. All the peripherals have DC and capacitance loading specifications, too, and adding a buffer may prevent a processor problem but leave a peripheral IC with a problem.

In the case where a buffer fixes a problem with the processor but leaves a peripheral with a problem, the bus may need to be split. This means that two or more separate data buses are needed, each with a separate buffer.

One simple way to split the bus is to have output-only drivers. This technique is useful if there are a large number of discrete output registers. All the registers are tied to one common bus, which is buffered from the processor bus with a unidirectional buffer. The processor sees only the load of the buffer, and the buffer is selected to be able to drive the register bank. The advantage to this method is that the buffer can be enabled all the time, eliminating the control logic.

Figure 2.21 shows a multichip design that uses a split bus and a unidirectional buffer. For fastest access, the EPROM and RAM are connected directly to the processor data and address buses. Low-drive peripherals are grouped with the EPROM and RAM on the processor bus. A second group of peripherals is connected to a second bus through a bidirectional buffer. A bank of registers is driven from a unidirectional buffer.

Regardless of what kind of buffers are used, the following rules must be obeyed:

- Adding buffers requires additional control logic to enable the buffers and control the direction of data flow. Be sure that the logic, especially if it's in

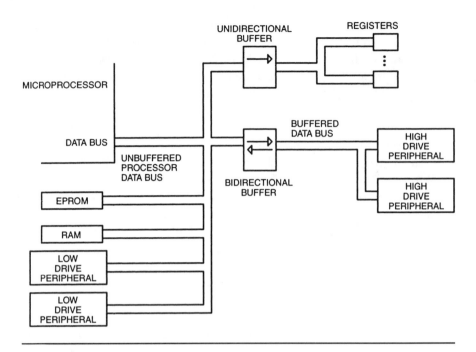

Figure 2.21
Buffering a Microprocessor Data Bus.

a PLD, has all the inputs needed to determine when to turn a buffer on and change the direction.

- Whether using one buffer or multiple buffers, be sure that the control logic allows each buffer to drive the bus only when the peripherals it controls are accessed. Simultaneously enabling two buffers causes bus contention, which can cause intermittent operation and even failure of the buffer ICs.

- Bus contention also can be caused if a buffer is enabled while the processor or a peripheral is driving the data bus. On a processor with a multiplexed data bus, driving the bus with a buffer while the processor is trying to latch a PROM address can be disastrous. Avoid that condition. Check the logic that enables the buffers to be sure they are not enabled at the wrong time, and check buffer output turnoff time to ensure that it is not too slow for the processor.

- The data bus has to propagate through the buffer, so add the propagation delay of the buffer to EPROM, RAM, and peripheral access time calculations. When using a buffer between the processor and a peripheral that requires write data to be stable at the *leading* edge of the write strobe, make sure that propagation delay through the buffer does not delay data far enough to cause problems.

Embedded Microprocessor Systems

Although usually fewer devices are tied directly to the processor address bus (especially with a multiplexed bus), the same considerations apply. Use buffers if the load exceeds the processor's capacity. One simplifying factor in these cases is that the address bus usually needs only unidirectional buffers.

Nonvolatile Memory

In many designs, it is necessary for the processor to remember certain parameters when the power is removed. Typical examples are the calibration parameters for some kinds of sensors, the enable/disable code for a burglar alarm, and the last channel selected on a television. In a multichip design, this can be accomplished by either nonvolatile memory, described earlier, or with an EEPROM. The EEPROM can be written by the processor but acts like a PROM, in that it remembers its contents when power is removed.

Some single-chip microcontrollers have built-in EEPROM just for these applications. However, general-purpose single-chip designs are at a disadvantage when nonvolatile storage, or any external peripheral for that matter, is required. Accessing conventional external memory uses up the I/O pins that are the primary reason for using a microcontroller in the first place. There are some standard interfaces that make not only nonvolatile memory but a number of other peripherals accessible to the designer using a microcontroller.

Figure 2.22 shows two of these interfaces: the Inter IC (IIC or I^{2}C) and Microwire.

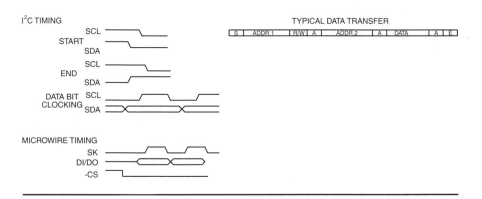

Figure 2.22
I^{2}C and Microwire Timing.

I²C Bus

The I²C bus is well suited to microcontroller applications. It uses two pins: SCL (SCLock) and SDA (SDAta). SCL is generated by the processor to clock data into and out of the peripheral device. SDA is a bidirectional line that serially transmits all data into and out of the peripheral. A microcontroller needs to supply only these two signals to communicate with any I²C peripheral. Several peripherals can share the same two-wire bus.

Since everything is communicated over two wires, the interface has every state and transition very well defined. For data transfers, the SDA signal is allowed to change only while SCL is in the low state. Transitions on the SDA line while SCL is high are interpreted as start and stop conditions.

If SDA goes low while SCL is high, all peripherals on the bus will interpret this as a START condition. SDA going high while SCL is high is a STOP or END condition.

Figure 2.22 illustrates a typical data transfer. The processor initiates the START condition, then sends ADDR 1. This is the peripheral address, which is 7 bits long and tells the devices on the bus which one is to be selected. Most I²C devices have address pins that are used to set part of the peripheral address. Next comes a single bit to select a read or write operation (1 for read, 0 for write).

After the read/write bit is sent, the processor programs the I/O pin connected to the SDA bit to be an input and clocks in an acknowledge bit. The selected peripheral will drive the SDA line low to indicate that it has received the address and read/write information.

After the acknowledge, the processor sends ADDR 2, which is the internal address within the peripheral that the processor wants to access. The length of this field varies with the peripheral. After ADDR 2 is another acknowledge, then the data are sent. For a write operation, the processor clocks out 8 data bits; for a read operation, the processor treats the SDA pin as an input and clocks in 8 bits. After the data comes another acknowledge.

Some peripherals permit multiple bytes to be read or written in one transfer. The processor repeats the data/acknowledge sequence until all the bytes are transferred.

A number of manufacturers, including Xicor and Philips, make EEPROM devices for the I²C bus. They have application notes that describe I²C solutions. Most microcontroller manufacturers have application notes that show how to interface I²C to their processors, including code. In addition to EEPROMs, Philips makes a number of I²C peripherals, including 8-bit port expanders codecs, and A/D converters.

EEPROMs, whether serial or conventional, have a limitation on the maximum number of write cycles that can be performed on the device. Early parts typically had a 10,000 write cycle limit. Newer parts allow around 1 million write cycles. But, even without this limitation, EEPROM write times are too slow for use as general-purpose RAM. If you need EEPROM, it will have to be in addition to, not in place of, general-purpose RAM. In using the serial EEPROMs, the simplest approach is to set aside a portion of RAM, load the EEPROM contents into it at power-up, and store data back to the EEPROM only if something changes.

One additional advantage of the serial EEPROMs is expandability. If you find sometime in the development cycle that more EEPROM is needed than originally planned for, just plug in a larger device. The pinouts are the same. But do not get extravagant. Typical serial EEPROM densities are 256 × 8, 1 K × 8, and so on.

As mentioned earlier, other peripherals are available with the I²C bus as well. Available parts include PLL (phase-locked loop) frequency synthesizers, analog-to-digital and digital-to-analog converters, and LED and LCD display drivers.

One drawback to the I²C bus is speed—the clock rate is limited to about 100 kHz. That limitation is not a severe speed penalty for a microcontroller that is toggling the lines in software, but faster interfaces are available.

Philips, which originally developed the I²C bus concept, also released a fast-mode I²C bus that operates to 400 Kbits/sec. In 1999, Philips announced a high-speed mode with operation to 3.4 Mbits/sec. High-speed and fast-mode devices are capable of operating in the older system as well, although older peripherals are not useable in a higher-speed system.

High-speed and fast-mode I²C also support a 10-bit address field, so up to 1024 addresses can be supported. Of course, to use the high-speed mode, you cannot control the interface using software, you need a processor that has a built-in I²C interface. Since the total capacitance on the bus can reach 400 pF, high-speed I²C requires active pull-ups, and fast-mode also requires active pull-ups if the total capacitance exceeds 200 pF.

I²C also supports a multimaster mode that is described in Chapter 7.

Microwire

Microwire is a three-wire serial interface used by National Semiconductor in its COPS processor family. The three signals are SI (serial input), SO (serial output), and SK (serial clock). SI and SO, respectively, are input to and output from the processor. The processor clocks data to the peripheral on SO and receives data on SI. Data in both directions is captured on the rising clock edge. Peripheral devices that transfer data in only one direction (such as display drivers that are only written, never read) may implement only one data line, SO or SI.

Unlike I²C, the Microwire protocol has no device addressing built into the serial bit stream. Microwire peripherals require a separate chip select input, one per device. This allows data to be transferred more quickly since address information is not needed. It requires more port bits, however, since one chip select, using one port bit, is needed per peripheral.

Each Microwire peripheral has a unique protocol based on the application. The number of bits and the meaning of each bit varies. National's Microwire EEPROMs, for example, have a 4-bit command followed by an address (7–12 bits, depending on memory size), followed by data (8 or 16 bits). The commands are erase, read, program, enable programming, and so on.

Microwire can transfer data faster than the original I²C, typically at MHz rates. The SPI bus, used by Motorola on its 68HC11 family, is similar to Microwire, and many peripheral ICs are specified as being compatible with either.

Both SPI and Microwire are implemented in their respective processors with hardware, which simplifies programming. However, peripherals using these buses can be interfaced to any general-purpose microcontroller using software-controlled I/O. Generally, the same types of peripherals available with the I²C interface also are available with SPI or Microwire. A summary comparison between SPI/Microwire and I²C is as follows:

Maximum bit rate:

Microwire: Into the MHz range.
I²C: About 100 kHz (standard), 400 kHz (fast mode), 3.4 MHz (high-speed mode).

Interface pins required:

Microwire: Three plus one chip select per peripheral.
I²C: Two, regardless of number of peripherals.

Number of devices sharing a bus:

Microwire: As many as there are chip selects available, as long as maximum loading is not exceeded.
I²C: Bus can address up to 127 peripherals.

Interface method:

Microwire: Usually dedicated hardware, can be implemented in software.

I²C: Software, but hardware ICs are available.

Other Serial Interfaces

Some manufacturers sell peripherals with a proprietary serial interface. Analog Devices, for example, has several A/D and D/A converters with simple serial data/clock schemes. These devices require three or more signals and can be interfaced to any general-purpose microcontroller.

DMA Timing Issues

Although DMA already has been discussed, the details of using DMA have not. One common mistake in designing with DMA is illustrated by the following scenario:

Processor 2 requests bus from processor 1.

Processor 1 gives up bus.

Processor 2 does whatever DMA thing it wants to do.

Processor 2 notifies processor 1 that DMA is done.

Processor 2 requests bus from processor 1 again. Sees bus acknowledge still asserted, takes bus.

Processor 1, still coming out of first DMA acknowledge state, takes the bus.

Bus contention or garbage data transfer results.

This is illustrated in Figure 2.23. The scenario can be avoided by not allowing the bus requester to request the bus a second time until the other processor has reacquired the bus after the first request.

Some processors, such as the 80186 and 386EX, have internal DMA controllers. These include address counters and logic to handle the DMA request inside the processor. On the 186 and 386EX, two pins are provided for peripherals to request a DMA transfer. The DMA controller requests the bus, does the transfer, and gives the bus back to the processor. The DMA controllers are programmable and can transfer data from memory to memory, memory to I/O, I/O to memory, and I/O to I/O. Most important, all the arbitration is handled inside the microprocessor, which saves on hardware and ensures that all the timing is correct.

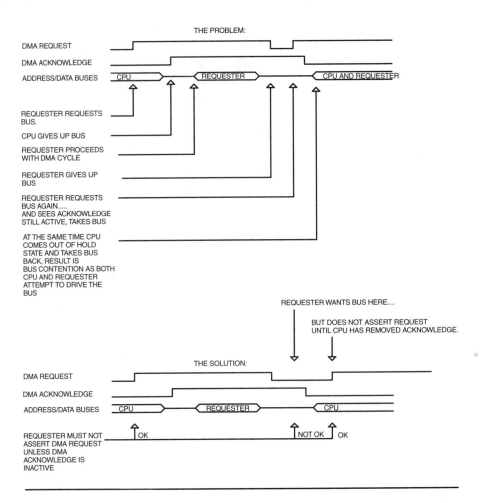

Figure 2.23
DMA Timing Problem.

Watchdog Timers

A microprocessor executes instructions from memory. If a nearby electrical discharge occurs, the processor data bus may be momentarily upset, and the processor can get a bad byte from the PROM. Or a software bug can result in a stack overflow, and the processor gets garbage when it tries to return from a subroutine. In either case, the processor will usually go "into the weeds," which is a shorthand way of saying that it will begin executing code in some unpredictable way, usually resulting in a system crash.

If this happens to the keyboard controller in your PC, you can just turn the power off and back on, and everything will be fine. If it happens to the processor that controls the rudder on a passenger jet, the results can be disastrous. To prevent this scenario, many embedded systems use a *watchdog timer* (WDT). The watchdog timer is a circuit that has to be triggered by the microprocessor on a regular basis. If that does not happen, the WDT resets the microprocessor. In most cases, if motors or other potentially dangerous equipment are connected, these are turned off at the same time.

The simplest WDT is a retriggerable monostable multivibrator, or one-shot. This flip-flop is latched by a trigger and stays in the latched state until some time has elapsed (determined by external timing components), then the output goes inactive. As long as the trigger keeps occurring before the circuit times out, the output stays active.

While many designers still design their own WDT circuits, several manufacturers make ICs that contain a WDT circuit. These parts also frequently contain other logic, such as power-on resets. Maxim, for example, makes a number of these parts.

Watchdog timers are straightforward to use. The time constant is usually around 0.5 to 2 sec. The WDT can be triggered by a port pin or a write to a particular address. The time constant can be a resistor-capacitor combination (on ordinary one-shots) or a digital delay from a constant clock.

One danger in using the WDT involves making sure the processor actually is running correctly. For example, a software bug may leave the processor executing a very tight loop, doing nothing but still servicing interrupts. If the WDT trigger is put in an interrupt routine, it does not generate a reset even though the processor essentially is locked up. If the WDT trigger is put in the polling loop, a software bug could disable interrupts, but the processor continues to go through the polling loop and still there is no WDT timeout.

In systems where safety or reliability concerns make it essential that the WDT reset the system any time a fault occurs, a more sophisticated WDT is needed. One method to make sure that both interrupts and the polling loop are running is to have one process set a flag location in memory each time that it executes (say, each time the interrupt occurs). The other routine (in our case, the polling loop) does not trigger the WDT unless the flag is set. Each time it *does* trigger the WDT, it resets the flag.

In systems where even this is not enough, a more sophisticated WDT can be designed, where each key process has to write a particular value to the hardware before the WDT will trigger.

Some microcontrollers, such as the PIC17Cxx series, have a built-in WDT. On some other processors, you can wire a timer to generate a reset when it

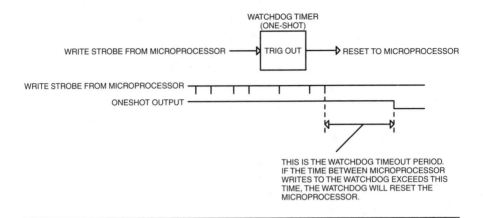

Figure 2.24
Watchdog Timing.

times out. The processor, instead of toggling a port or an I/O strobe, resets the timer count periodically.

Figure 2.24 shows the basic operation of a watchdog timer. Note that the write strobes from the microprocessor need not be evenly spaced, as long as they are always shorter than the timeout period. The timer in the figure is shown as a simple block; it could be implemented as a one-shot IC, a digital divider, or as part of an off-the-shelf IC that includes other supervisory functions, such as a power-up reset. Whatever method is used, the watchdog needs to remove the reset output once the processor has been reset, or else the processor will be held in reset forever.

In-Circuit Programming

As mentioned in Chapter 1, sometimes the ability to reprogram the memory in-circuit is a useful feature. If you are using flash memory external to the microprocessor, in-circuit programming is fairly straightforward. You can treat the flash like a slow RAM. Usually a sequence of data writes are required to enable programming. You usually want to use a memory device with block erase, so you can leave the programming code in one portion of the memory while reprogramming the rest of the device.

If you are using a microcontroller with internal flash memory, some extra considerations are required. The device pins needed for programming typically are shared with other functions. As an example, look at the Microchip PIC 16F84. The 16F84 has 1 K of internal flash memory. To program the device, data are loaded and read serially using two of the port pins, RB6 and RB7. The MCLR pin is used to provide the +12 V programming voltage.

Embedded Microprocessor Systems

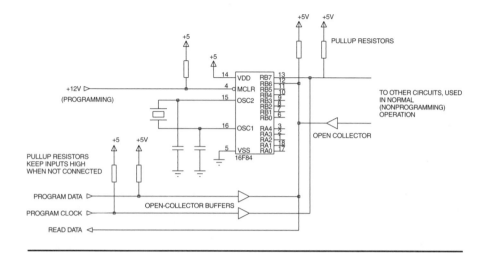

Figure 2.25
In-Circuit Programming of a PIC16F84.

The problem with this approach is that the 16F84 has only 18 pins; in many designs you would need RB6 and RB7 for some other function. So you have to make these pins do double duty, functioning as normal I/O pins and also as in-circuit programming pins. This is complicated by the fact that RB7 is bidirectional in programming mode.

Figure 2.25 shows one way to implement in-circuit programming with the 16F84. The two programming pins are driven with open-collector buffers. When not programming, the inputs to the open-collector buffers are high, making the outputs float. The pins then can be driven by (or drive) external circuitry. For programming, the external logic must be turned off and the programming buffers drive the pins. The external device doing the programming must have separate data in and data out pins to make this work. In the figure, RB6 is an input in normal operation, RB7 is an output. Of course, you have to be sure that the external logic is not confused when programming data is applied to the device pins.

An alternative way to do this is to connect RB6 and RB7 through a PLD and switch the programming function on and off with a control input.

The +12V for programming comes from an external supply that can be switched off or switched between +12V and +5V.

This example is specific to the PIC processors, but you find similar situations with other processors that have flash memory and in-circuit programming capability. The NEC μPD70F300 series processors allows you to use either the asynchronous serial input or a three-wire synchronous serial interface for in-circuit programming. Either way, you have to do something similar to the PIC arrangement if these functions are used on the board in normal operation.

Internal Peripherals

A number of processors and microcontrollers have built-in peripherals. These usually are the peripherals commonly used in many embedded designs. Nearly every processor intended specifically for embedded systems includes at least one timer. Other common peripherals include serial ports, DMA controllers, watchdog timers, and interrupt control functions. Be sure you take into account specific restrictions of the peripheral device in the microprocessor you are using.

Take a look at the Atmel AT90S4414 processor. This device includes an 8-bit timer and a 16-bit timer. Both timers include a prescaler that can divide the input signal by 8, 64, 256, or 1024. Both timers can be clocked from the internal processor clock or from an external pin. When using the external clock input, the signal is internally synchronized with the CPU clock. So the maximum frequency you can input to these timers is about half the CPU clock. If you are using an 8 MHz 90S4414, the maximum timer frequency is a little less than 4 MHz. If you're using an 8 MHz 90S4414 but running it at 6 MHz, the maximum timer input frequency is about 3 MHz.

Contrast that with the Microchip PIC16C62. The PIC16C62 also has timers that can operate from an external clock. The internal clock on the PIC devices is one quarter the external input, so it would appear that the PIC clock rate (20 MHz in/4 MHz CPU) limits you to a slower input clock rate than the fastest Atmel AT90S clock rate (8 MHz). If the external clock input is synchronized to the CPU clock, that is true. But the PIC devices have a mode where the timer clock need not be synchronized to the internal CPU clock. In this mode, the clock input frequency can be as high as 50 MHz. However, in this mode, the timer cannot be used for any operations that require synchronization, such as PWM or capture/compare.

The 80186/80188 DMA controllers, already mentioned, have two modes of synchronization: source and destination. Source synchronization is intended for use when the DMA source generates the request, wanting to send something to the destination. Destination synchronization is used when the destination device wants to get something from the source. In practice, either device can be the requestor, since source/destination synchronization is programmed into the DMA controller by the firmware. However, the choice of source versus destination synchronization affects the timing. Source synchronization permits faster transfers but requires that the DMA request be removed before the end of the DMA write cycle. Destination synchronization is slower, but has more relaxed timing for removal of the DMA request.

Embedded Microprocessor Systems

Most embedded processors include pins that can be used as external edge-sensitive interrupts. Like the timers, these usually are synchronized to the internal processor clock, which limits the minimum pulse-width that will be reliably recognized as an interrupt request.

Design Shortcuts

It sometimes is possible to simplify a design by taking some shortcuts with the hardware. A few of these are described in this section.

Partial Address Decoding

Say a microprocessor with a 20-bit address bus (1 Mbit space) needs an 8 K × 8 EPROM at location FE000. Decoding the entire range of addresses would require that 6 bits (A14–A19) be decoded. If only A16–A19 is decoded using a four-input NAND gate, the EPROM will be addressed anytime that the processor accesses anything in the range F0000–FFFFF. This works as long as nothing else needs to go in that range. The EPROM may be accessed in the address space starting at F0000, F2000, F4000, and so on.

Linear Address Decoding

Assume that a microprocessor with a 16-bit address bus (64 K space) needs an 8 K EPROM and a 2 K RAM. The EPROM goes at location 0. To decode this, connect A15 from the processor to the EPROM -CE input. Connect A14 through an inverter to the RAM -CE input. Now the EPROM is accessed anytime that A15 is 0, which is anywhere in the lower 32 K. The RAM is selected any time that A14 is 1, which occurs from 4000 to 7FFF and from C000 to FFFF. The first range also enables the EPROM, causing a bus conflict. But, if the software addresses the EPROM from 0000 to 1FFF and the RAM from C000 to C7FF, the EPROM will be deselected and there is no bus conflict. This principle can be expanded to as many decodes as there are available address lines.

Buffer Always Enabled

When using data bus buffers, it is not always necessary to enable and disable the buffer's tristate outputs. Instead, the buffer can be enabled all the time, usually by grounding the enable pin, and the direction can be controlled. In this scheme, the processor side of the buffer normally is the input and the peripheral side normally is the output. The direction is reversed only when the processor reads from the peripheral. When using this technique, the buffer has to switch directions to drive the processor data bus only when the processor is not driving it and must switch back only after the peripheral has stopped driving it. Otherwise, you will get bus contention.

EMC Considerations

Most embedded systems end up in products that require certification to EMC standards. In the United States, the Federal Communications Commission has limits on how much RF energy a product can emit. The European community also has standards for EMC compatibility, and they include susceptibility to external RF fields and to ESD (electrostatic discharge).

Entire books have been written on the subject of designing for EMC, so here we concentrate only on those aspects of EMC design that bear directly on embedded systems.

The first consideration for EMC design is limiting RF emissions. Since microprocessors use crystals and those crystals operate at RF frequencies, an embedded system radiates at the processor frequency. Embedded systems are digital, so there usually are emissions at the odd harmonics of the processor crystal frequency. In addition, regular signals such as ALE or address lines can radiate at some frequency other than the processor clock frequency.

RF energy can be radiated from PC board traces and wires that interconnect the system. Multiprocessor systems that have more than one processor operating at the same frequency are a particular problem, because usually at some point in the test the power from the oscillators will sum, causing considerable energy to be radiated.

Controlling EMC Emissions

The following are a few guidelines for controlling EMC emissions from your system.

Put a small (50–75 ohm) resistor in series with oscillator outputs and signal lines with more-or-less regular signals, such as ALE. This both matches the output to the PC board, reducing ringing, and dampens the rise time of the waveform, reducing the effect of the odd-order harmonics.

Put board-mounted EMC filters on each I/O line. Of course, if you have a very fast interface (such as video or 100 MHz Ethernet), you cannot do this, as it will affect the signals you want to have.

Shield the processor board and all interconnected electronics. Sandwich clock lines between the power and ground planes.

In multiprocessor systems, if you have multiple processors on a single board, all operating at the same frequency, do not give each processor a separate oscillator. Have a single oscillator and distribute it to the various processors. If your multiple processors are on different boards (or you cannot distribute a single clock for some reason), see if you can stagger the oscillator frequencies slightly. For instance, instead of running all the processors at 20 MHz, run one at 19.966 MHz, one at 20 MHz, and one at 20.0333 MHz. This will push the third harmonics apart by 100 kHz.

Pick a processor with lower EMC emissions. To run a Microchip PIC processor at 5 MHz, you have to put in a 20 MHz clock since the PIC divides the clock internally by 4. The third harmonic of a square wave usually contains considerable energy, and the third harmonic of 20 MHz is 60 MHz, right in the worst part of the radiated emissions test spectrum. On the other hand, if you use a processor in the Atmel AT90S family, you can run at 5 MHz with a 5 MHz input—there is no internal clock division. This clock rate will have lower radiated emissions in the spectrum that is tested for radiated emissions. Of course, you do not want to let EMC considerations completely drive the design, but if it matters little otherwise, you might think about EMC in choosing a processor.

ESD Protection

Protecting a system against ESD involves many of the same techniques used for preventing RF emissions problems. ESD interference often takes the form of RF energy, and the same things that keep RF in a box tend to keep it out.

Shielding. Shield the system and use shielded cables where appropriate.

Filters. EMI filters on signal lines will help keep ESD off the processor board.

Grounding. Make sure ESD energy has a low-impedance path from the discharge point (usually on the chassis or other operator-accessible areas) to ground. If the lowest-impedance path for ESD is through the ground plane on your board, that is where it will go. Avoid having ground loops through your board wherever possible. Do not ground the CPU board to the chassis through the mounting standoffs. Instead, have a single wire return to the DC power supply and have a single point connection to chassis there.

Interfaces. Unfortunately, the embedded system often has to talk to other devices. The interfaces often require a ground reference at both ends to operate correctly, as for example, RS-232. Consequently, you are forced to design in ground loops just because of the interface requirements. You may need ferrite beads or EMC filters where the interface signals enter and leave the board. This will attenuate the high-frequency energy of an ESD pulse.

Watchdog timer. Add a watchdog timer to your circuit so, if ESD corrupts program execution, the system can recover.

Isolation. Use optical isolation where it makes sense, especially if you are connecting to a system with more severe ESD requirements than your own.

Other EMI Considerations

Ground Loops   It is increasingly common for embedded systems to be controlled from an external computer. If the external computer is connected to a different AC power source than the embedded system (such as a 120V computer connected to a 280V, three-phase machine), you may get ground loops between the two systems. Be sure the grounds are common. If you cannot make the grounds the same (maybe because the customer controls where the computer plugs in), optically isolate the interface.

Differential Interfaces   Differential interfaces, like RS-485 or LVDS, can reduce susceptibility to ground noise and other types of electrical noise. But, while differential interfaces are good at noise rejection, they have to be built from real parts and those parts usually have a maximum common-mode offset voltage that differential receivers can tolerate. An RS-485 interface with correct *differential* voltages but with a 20V ground offset between the two systems is not going to work. And remember that two grounds that are the same most of the time are not necessarily the same all the time. I have seen interface drivers and receivers actually destroyed when an air conditioner switched on, yanking the ground on an embedded system many volts away from the ground of the system it was communicating with.

Radiated Susceptibility   Interference from external RF sources such as cellular telephones and walkie-talkies can affect an embedded system. The interference may directly affect the processor circuitry or cause problems through secondary effects.

To minimize the possibility of susceptibility, use small value pull-ups on unterminated lines. A 100K pull-up on a CMOS input makes the input impedance about 100K and potentially capable of picking up strong RF signals. Use 10K or add an R/C terminator to the signal.

A sensor such as a strain gauge may pick up the RF and produce erroneous outputs. Protect against this by performing sanity checks in the software and protect against continuous interrupts from a berserk input that is picking up RF.

Since embedded systems usually control something in the real world and those things often involve motors, be sure you do not build a self-destroying system. A rotating, insulated pulley, driving an insulated belt, can become a fairly good static electricity generator. I saw a system once with an insulated plastic drum running against a mylar strip that could create a half-inch arc—not good for the microprocessor that was controlling everything.

In a system with DC motors, it is a good idea to have a separate return path for the motor voltage back to the power supply. This prevents the startup and braking surge currents in the motor from yanking the ground on the electronics boards around.

Finally, in designing for EMC, it is a good idea to keep in mind what a certification engineer I know used to say, "First, it has to work." In other words, it is not a good idea to make changes or design compromises that let the system pass the EMC tests but degrade performance in the real application. I have seen so much inductance added to a clock line in an attempt to reduce emissions that the circuitry receiving the signal could not operate reliably. I have seen video cables in a high-speed imaging system with huge ferrite beads added. This fixed the emissions problem but killed the video signal. First, it has to work.

Microprocessor Clocks

Crystals and ceramic resonators were mentioned earlier in the chapter. The selection of a crystal reference for a microprocessor often seems to be a source of mystery, so this section will try to clear it up a little.

Most microcontrollers and many microprocessors have an internal oscillator, like that shown in Figure 2.26. This usually is a high-gain, inverting amplifier stage. The crystal is connected with external capacitors C1 and C2 to make an oscillator. Figure 2.26 also shows the oscillator circuit with the equivalent circuit of the crystal. At resonance, the crystal looks like a series L-C circuit with some series resistance and some parallel capacitance. A crystal oscillator of this type is called a *Pierce oscillator*. A Pierce oscillator always will resonate at the fundamental frequency of the crystal, unless external components are added to force operation at a different frequency.

Load Capacitance

Capacitors C1 and C2 are needed to allow the oscillator to start. C1 and C2 are typically between 20 pf and 100 pf. Parallel resonant crystals are specified with a particular load capacitance. Ideally, the load capacitance of the circuit will match the specified load capacitance of the crystal. The circuit load capacitance is given by the equation:

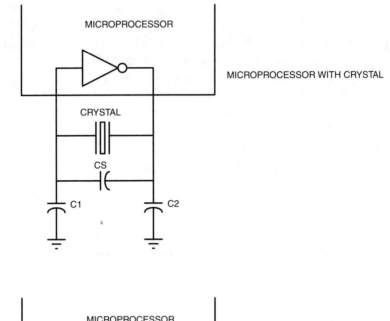

MICROPROCESSOR

MICROPROCESSOR WITH CRYSTAL

CRYSTAL

CS

C1   C2

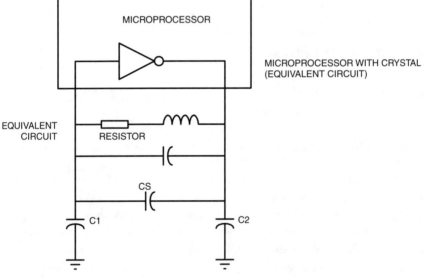

MICROPROCESSOR

MICROPROCESSOR WITH CRYSTAL
(EQUIVALENT CIRCUIT)

EQUIVALENT
CIRCUIT   RESISTOR

CS

C1   C2

Figure 2.26
Crystal Oscillator.

$$Cload = \frac{C1C2}{C1+C2} + Cs$$

Cs is the stray capacitance in the circuit, usually around 5 pF. So if you have a crystal that is specified with a load capacitance of 30 pF, C1 and C2 each would be about 50 pF. However, the optimum values for C1 and C2 are a trade-off between frequency stability and startup time. If C1 and C2 are too large, the oscillator will not start. If they are too small, the oscillator theoretically will not start, although the circuit capacitance often is enough to make it work anyway.

Series versus Parallel

Crystals can oscillate at two frequencies, series or parallel. A series crystal is intended for use in a circuit with no reactive components (no C1 or C2), while a parallel crystal is intended for use in a circuit with these components. For the purposes of embedded applications, the difference is that the resonant frequency of a parallel crystal is achieved when the crystal is installed in an inverting circuit, like that shown in the figure. A series crystal is used in a circuit with a noninverting amplifier. A series crystal can be used in a parallel circuit, and vice-versa, but the frequency will be off by about 0.02 percent.

The stability of a Pierce crystal oscillator usually is in the range of 0.1 percent. This is accurate enough for many applications, but if you are designing something that has to keep track of the time of day, this will not be close enough. A 0.1 percent error is a drift of almost 1.5 minutes per day. For applications that are extremely sensitive to frequency, you will need a very accurate oscillator design, an external oscillator, or a means to adjust the crystal frequency.

Fundamental Mode

I got a call from the factory one day. The company had a batch of processor boards that worked, but everything worked funny. The response was slow, and the machine would not function correctly, although all the operator screens looked okay. It turned out that the 24 MHz crystals were overtone crystals. In our circuit, they operated at their fundamental frequency of 8 MHz. Make sure you specify the right crystals and the right circuit around the crystals. This most likely is a problem with AT-cut crystals in the 20–30 MHz range, where you can get both types. Below 20 MHz, most crystals are fundamental; and above 30 MHz, most crystals are overtone. This is because the crystal blank gets very thin when you try to make a high-frequency, fundamental-mode crystal.

Ceramic Resonators

A ceramic resonator is less expensive than a crystal but less accurate as well. Ceramic resonators typically have a frequency accuracy about one tenth that of crystals.

External Oscillators

The two considerations for using an external oscillator are the drive level and the processor connection. Most modern microprocessors have CMOS technology, so an oscillator with CMOS drive levels must be used. Since most microcontrollers provide two pins for a crystal, you drive only one when using an external oscillator. Make sure you drive the right one, and terminate the other as specified by the manufacturer. Some processors need the unused pin to be floating, some want it tied low, and so on.

Internal A/D Converters

Many microprocessors include an internal analog-to-digital converter. The Microchip PIC16C7x series is a typical example. The parts in this family have an internal 8-bit A/D converter. The parts have from four to eight analog inputs. There is only one A/D converter, but an internal analog multiplexer allows the A/D converter to process any of the inputs, one at a time. Any microprocessor design that uses an A/D converter, whether internal or external, has to take into account some considerations.

Input Multiplexing

The Microchip A/D converter handles multiple inputs by selecting one at a time under software control. Once an input is selected, a settling time must elapse before the A/D conversion can start. If the conversion is started immediately, the result will be incorrect. The software must take this delay into account.

Reference Voltage

The Microchip devices allow you to use one input pin as a reference voltage. This normally is tied to some kind of precision reference. The value read from the A/D converter after a conversion is

$$\text{Digital word} = (\text{Vin}/\text{Vref}) \times 255$$

The Microchip parts also permit the reference voltage to be set internally to the supply voltage, which permits the reference input pin to be another analog input. In a 5V system, this means that Vref is 5V. So measuring a 3V signal would produce the following result:

$$\text{Digital word} = (\text{Vin}/\text{Vref}) \times 255 = (3\,\text{V}/5\,\text{V}) \times 255 = 153_{10} = 99_{16}$$

However, the result depends on the value of the 5V supply. If the supply voltage is high by 1 percent, it has a value of 5.05V. Now the value of the A/D conversion will be

$$(3\,\text{V}/5.05\,\text{V}) \times 255 = 151_{10} = 97_{16}$$

So a 1 percent change in the supply voltage causes the conversion result to change by two counts. Typical power supplies can vary by 2 or 3 percent, so power supply variations can have a significant effect on the results. The power supply output can vary with loading, especially if there is any significant drop in the cabling that connects the power supply to the microprocessor board. So, if your design needs all the analog inputs and cannot use an external reference, be sure power supply variations cause no accuracy problems.

Hardware Checklist

This is a summary of the information scattered through the chapter.

- Verify EPROM, RAM, and peripheral access times. Add wait states if necessary.
- Verify that setup and hold times are met, both to the peripheral device and to the microprocessor.
- Make sure the crystal type matches the system specifications. If using an external oscillator, be sure it matches duty cycle and drive-level requirements of the microprocessor.
- Verify that bus loading is not exceeded.
- Verify that loading of I/O port pins is not exceeded—either sinking or sourcing current.
- Verify that inputs to timers meet frequency and duty cycle limitations of the part. This usually depends on the microprocessor or timer IC clock speed.
- If bus buffers are needed, make sure that each buffer is enabled only when the peripherals it is accessing are selected. Watch for bus conflicts from turning the buffer on too early or turning it off too late. Account for buffer delays in timing calculations.
- Avoid bus contention between peripherals.

- Make sure wait state timing gets wait requests asserted in time.
- Make sure that the WDT functionality matches system requirements.
- Make sure the internal peripherals are compatible with your requirements.
- Make sure the DMA controller logic can handle back-to-back accesses.
- Make sure you take EMC requirements into consideration.
- Be sure ADC references and input ranges give the right resolution and accuracy.

Example System

Appendix A contains the schematic of the pool pump timer system. An 80C31 microcontroller is used, with external ROM and I/O ports. Since there is only one read and one write port, no address decoding is needed. The read buffer (a 74HC244) is directly enabled by -RD and the write register is clocked by -WR. A 74LS123 one-shot provides a WDT. The user key inputs connect to J2 and are switch closures to ground. Switch inputs are debounced in software by the 250 Hz interrupt code. The display, not shown, consists of four seven-segment displays and three high-intensity LEDs. The display is multiplexed in software, so only one seven-segment drive register is required. Input power and the pump relay coil are connected via a four-terminal barrier strip.

Hardware Specifications Outline

The following is a generic outline for the hardware specifications.

Overview. A brief description of what the outline covers.

Related Documents. ANSI, IEEE specs. It also may include a reference to the product requirements document.

Description. A brief description of the hardware and what it does. For example, The xyz board controls the three-axis robotic motors for the robotic arm. This includes the stepper and DC servo control and the related limit switches. The xyz board is controlled by a 10 MHz 80C188 microprocessor. DC servo motion is implemented using the PID filter algorithms developed by the Motion control research group.

External Connections. Describe what the board connects to or controls:

X- and Z-axis DC servo motors and encoders, and Y-axis stepper motor. RS-232 interface to system control computer.

Limit switch inputs (3).

Common emergency stop switch.

Description. This describes the circuitry and what it does: The xyz board is implemented with an 80C188 operating at 10 MHz (20 MHz input). DC motors (X- and Z-axis) are controlled by LM629 motion control ICs. The Y-axis stepper is controlled with the timer 1 output of the 80C188. Limit and emergency stop switches are read via an 8-bit status register. An 8-bit command register provides discrete control bits, including stepper direction. The RS-232 interface to the system controller is implemented with a 16550 UART.

Software Interface. This is where you tell the software engineers everything they need to know to use the board.

Memory:
    256 K flash, selected with 80C188 UCS signal.
    256 K RAM, selected with 80C188 LCS signal.
    1 K EEPROM, decoded at 3000h (h = hexadecimal).
I/O: 00
Read: Status register:
    Bit 0: Emergency stop switch (1 = stop)
    Bit 1: X-axis limit switch (0 = limit reached)
    Bit 2: Y-axis limit switch (0 = limit reached)
    Bit 3: Z-axis limit switch (0 = limit reached)
    Bits 4–7: Unused
Write: Command register:
    Bit 0: Y-axis stepper direction (1 = forward)
    Bit 1: X-axis brake input to power driver (1 = brake on)
    Bit 2: Z-axis brake input to power driver (1 = brake on)
    Bits 3–7: Unused
02, 03: X-axis LM629
04, 05: Z-axis LM629
06, 07: 16550 UART
Interrupt usage: Describe how the interrupt inputs are connected:
    INT0: X-axis LM629 interrupt
    INT1: Z-axis LM629 interrupt
    INT2: 16550 interrupt
    INT3: Unused
Port bits: Does not apply to the example used here, but in a micro-controller design, you would describe the usage of the port bits:
    RB0: Interrupt from external pushbutton
    RB1: Turns on diagnostic LED (0 = LED on)

RB2-RB7: Unused

RC0-RC7: Communication FIFO data bus

This description also would apply to a multichip design that used a PIO chip or had discrete I/O ports implemented in a PLD. Similar descriptions are needed for other software-controlled functions. For instance, if the design includes analog-to-digital or digital-to-analog converters, specify the range in bits and how it corresponds to what is being measured controlled. For example, an 8-bit ADC might be specified as representing 100 psi at full scale.

Interface Protocol. If the board communicates with an external system, this describes the protocol used.

Appendix. If there are calculations that went into the design, such as required motor torque or interface throughput requirements, these are collected into an appendix.

This chapter has discussed numerous ICs that can be used to interface to microprocessors. The electronics world seems to be defined by shorter product lives, bigger memories, and faster processors. Although the specific techniques described here may be implemented in a CPLD, FPGA, or ASIC and access times get ever shorter, the basic principles still apply.

In Chapter 3, we look at the software side of the design process.

Software Design 3

Some will be surprised that this chapter is shorter than the preceding one. This one chapter cannot be a complete course on software development, any more than the hardware chapter can be a course on logic design. I originally planned to cover here only those aspects of software development that relate to embedded systems. I finally decided on a compromise of showing some basic concepts, while focusing on the embedded environment. I have tried to leave general software concepts to the books written for that purpose, except where those concepts have a unique bearing on embedded design.

As mentioned in Chapter 1, software for embedded systems has different requirements from software running on a PC or workstation. The users of a computer will not even notice if your software takes a few tens of milliseconds to respond when they press a key. But if the electronic ignition in their car delays firing the spark plugs by the same amount, they will be very upset as the car stutters to a stop alongside the interstate. Embedded systems work in real time.

After our users get towed home, they turn on their microwave. They do not want to see a message telling them that thawing the meat requires installation of an additional 32 MB of memory. Embedded systems are self-contained, with memory and other resource limitations. Software for embedded systems has to work within these constraints, and that is what sets it apart from the software in a PC.

As mentioned in Chapter 1, there are a number of ways to document software for an embedded system. The one unforgivable crime is to leave undocumented code behind you. The level of documentation required depends on the customer and the complexity of the system.

Before going into the various documentation methods, I want to give you some background on the pool timer software, which will be used for illustration throughout this chapter. The pool timer software has a polling loop and a chunk of code that processes the 250 Hz timer interrupt. The timer code will be examined more closely in the chapter on interrupts. For now, think of it as a "black box" that passes certain information to the rest of the code.

The polling loop exists in some form in nearly all embedded systems (systems that use an RTOS are an exception and will be covered in a later chapter). Sometimes called an *idle loop*, *superloop*, or *background loop*, the polling loop is where the software spends its time when not processing interrupts. The polling loop determines how the various tasks are scheduled and executed. While everything in an embedded system usually works in real time, the polling loop holds those tasks that do not need immediate attention. An example of this in the pool timer is the code to handle the push buttons. The timer code tells the polling loop when the button is pressed. The polling loop may be slow getting to it, but not slow enough for the user to notice. In a completely interrupt-driven system, the polling loop may be a one-instruction jump to itself. (I did this once with a digital signal processor [DSP] design.) But in most systems, the polling code does some actual processing.

The pool timer polling loop continuously checks these processes:

- The motor control code, which handles motor on/off control.
- The powerfail code blinks the display after power-up until the user presses the SET button.
- The normal timekeeping code keeps track of when the time changes from ON to OFF and handles on/off/override.
- The time set mode allows a new ON or OFF time to be entered.

The polling loop sequentially checks for each event (timeout, button press, etc.) and takes the necessary action.

Data Flow Diagram

There are a number of ways to describe the software design, depending on what information is to be conveyed. One method is a *data flow diagram*. The data flow diagram shows each process as a block (or circle). Lines connect the blocks, showing what information is passed between the processes.

Figure 3.1 shows a data flow diagram for the pool timer. The timer interrupt tells the motor control code if the water level is low and tells the other three processes if the SET button is pressed. The time set code passes updated time back to the normal timekeeping code so that time rollovers will result in the correct initial time. The other paths can be seen on the diagram.

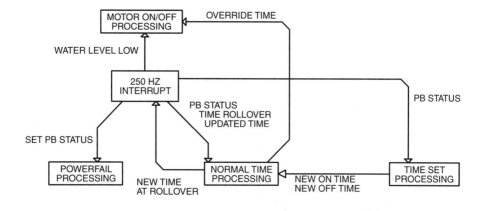

Figure 3.1
Pool Timer Software Data Flow Diagram.

Data passed to and from the hardware is not shown on this diagram, but if this were, a box would represent the keypad with push-button information going into the interrupt code and another box would represent the display accepting time data from the interrupt code. Some engineers draw the hardware as just another process, others denote it with a special symbol.

State Diagram

Figure 3.2 shows a *state diagram* for the pool timer. The state diagram shows each possible state for the software and what inputs cause a change to another state. As the figure illustrates, the only user input that changes modes of operation is the SET button. In a more complex system, each input that can cause a state change is shown, pointing from the old to the new state.

Figure 3.2 also shows the states within the time set state. When setting on time, pressing the OFF button switches to off time set, and vice versa. Pressing SET returns the timer to the normal timekeeping state. Although not shown here, you can add comments under the state blocks, showing what each state does.

The pool timer software is simple enough that the details of the time set and normal timekeeping states could be shown as part of the overall state diagram. In a more complex system, each state may require a page or more to describe, and several parallel processes can be going on at once, each in a different state.

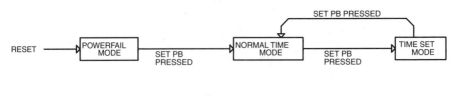

OVERALL STATE DIAGRAM

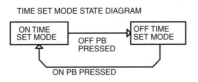

TIME SET MODE STATE DIAGRAM

Figure 3.2
Pool Timer State Diagram.

Pseudocode

A third method of documenting the software, and one that I prefer, is to use pseudocode. Appendix A shows two listings. The first is a high-level functional or logical description. This listing describes in concise English exactly what the software is going to do. I like it because it is a good way to see if I understand everything I need to know before coding.

The next level down is actual pseudocode, also shown in Appendix A. The pseudocode is still a description in structured English, but additional text describes what flags, variables, and other elements are manipulated to implement the described functions. The advantage of this method of documentation is that the functional description can be written, with details filled in as the design progresses, eventually turning into pseudocode. The pseudocode and description then become detailed comments for the actual code.

Functional descriptions and pseudocode are useful for simple systems written by a single programmer. If multiple programmers will work on a project, and especially if multiple processors will be used, then some means *must* be implemented to fully describe what information is passed between the processes and/or functions done by each programmer. Timing information for real-time functions is crucial. For example, function X must be called within Y milliseconds of function Z or the stepper motor will lose sync. In a

Embedded Microprocessor Systems

complex system, it is a good idea to have multiple levels of documentation. An overall block diagram shows what data are transferred in and out of the system, another diagram shows what is passed between subsystems or boards or the major firmware functions, and psuedocode or state diagrams describe how each of the functions works.

The following pseudocode is a description of the software I implemented in a simple protocol converter. The system accepted serial RS-232 data from a host system, did some processing, buffered the data, and sent it to a second system via a different interface. XON/XOFF protocol was used to control the data flow:

```
If serial data available, read the data,
        do some proprietary stuff I cannot reveal, then
        store the data in a first in, first out (FIFO) buffer.
If the buffer gets too full, and if XOFF was not sent yet,
        send XOFF to the host.
If there are data in the FIFO buffer, and if the output interface
        is ready, send a byte from the FIFO buffer to the output.
If the buffer gets close to empty, and if XON was not sent
        yet, send XON to the host.
```

The software just went around and around this loop. The actual system did some error checking and other tasks that I left out to simplify the description. Note that the input and output processes are mostly independent of each other. The input stores data in the FIFO buffer without knowing what the output is doing. Similarly, the output process sends whatever is in the FIFO buffer if the output device is ready for it. If the output is not ready, the code just waits until the next pass through the loop and checks again. The only way one process knows what the other is doing is by how full the FIFO buffer gets. This is a simple example that illustrates processes that are parallel and (mostly) independent.

The more independent processes are (that is, the less data they share in common), the more predictable each piece of code is. If processes are completely independent, some unexpected flaw in one process will not cause an unexplained failure in another. But there are exceptions to this, which we will look at later.

Flowcharts

The last documentation in Appendix A is the flowchart for the pool timer. To save space, only the polling loop is shown. Flowcharts seemed (to me, anyway) to lose favor for a while, but they are coming back into popularity since tools now simplify the process of creating and maintaining them. One problem with using flowcharts to document code is the difficulty of showing the effects of interrupts. Flowcharts tend not to be maintained with the code, so they often get out-of-date as the code changes (in fact, this tends to be a problem with any graphical representation, including state diagrams). Still, the flowchart is a good graphical way to show the execution of sequential code.

Partitioning the Code

The charts and diagrams shown so far assume that you already know how the code will be functionally partitioned. The process for determining this break-down and developing the code is the same as for any software, but with a few additional considerations for embedded systems:

- In a PC, an operating system controls access to the disk drive, display, and other peripherals. While there are real-time operating systems, which will be covered in Chapter 8, most simple embedded designs have none. In these cases, some mechanism must arbitrate access to peripherals and memory. Two serial transmit routines cannot both be filling the same buffer, for example. The simplest way to do this is to have each resource (serial I/O, interface buffers, etc.) controlled by only one piece of code. This seems obvious, but it is easy to cheat when sending a byte over the serial interface is as simple as an assembly language move instruction. In cases where this rule cannot be followed, usually because of throughput, make sure that conflicts do not occur.
- Since the code is stored in programmable read-only memory (PROM), self-modifying code is out. This is considered bad practice anyway, but it is nearly impossible in an embedded system. *Self-adapting* code, however, is possible if nonvolatile storage is available.
- Some software engineers write code for maximum maintainability, some for maximum efficiency, some for minimum space. Some embedded systems can be written only for speed, or they will not work. Many embedded systems cannot tolerate a function that just "goes away" for a long time, say, to sort data in a table. The definition of a *long time* depends on the system, but may range from seconds down to milliseconds (even microseconds in a DSP-based design).
- An embedded system is not a general-purpose computer. There often is no display to handle errors. There may be no human within miles, so "Hit Any Key to Continue" generally is not an option. Errors must be handled and often operation must resume after they occur.

Embedded Microprocessor Systems

- A real-time system handles asynchronous external events. This means that a switch can be closed at any time and an interrupt can occur between any pair of instructions. The code must handle all timing combinations without error. This will be discussed further, with examples, in Chapter 4.
- There is no nanny to be sure the hardware is in a known state. The embedded code must initialize *everything* at power-up. More about this at the end of the chapter.

Other considerations include:

Safety. If the code in a PC goes "off into the weeds," the disk drive spindle motor cannot reach out and grab the user's tie. In some real-time systems, an unsafe scenario is a real possibility. When in doubt, *make it safe*. Turn the motors off. Shut down the high voltage.

Hardware damage. This is a less serious version of the safety issue. An example is control of a direct current (DC) or stepper motor. If the software directly controls all four transistors in a motor H-bridge, it can turn on the wrong pair and destroy the transistors. Situations like this must be dealt with in embedded systems.

Mechanical delays. In a PC, the operating system takes care of the delay between turning on a disk spindle motor and waiting for the disk to come up to speed. Similarly, embedded system software must take into account the fact that mechanical systems are often much slower than the processor. A real-world example involves AC motors controlled by two relays. A run relay provides AC for normal operation. For faster stopping, a second relay provides a momentary pulse of DC to brake the motors. Relays have a delay of many milliseconds from when voltage is applied to the coil until the contacts close. There is a similar (usually longer) delay from the time the coil voltage is removed until the contacts open. So when switching from running to braking, the software has to introduce a delay between turning off the run relay and turning on the brake relay. Without the delay, the run relay will fail in a quite spectacular manner, accompanied by blue sparks and smoke. A second delay is required to inhibit run startup after the brake relay is opened. Since the braking relay is a large contactor with an opening time in the tens of milliseconds, the second delay has to be much longer than the first.

Recovery time. Many peripheral ICs have a recovery time. You cannot perform a read or write cycle until so many clocks have elapsed since the last cycle. Be sure to check for and abide by these. If you do not, the resulting problems can be intermittent and difficult to find.

EMI. Although EMI considerations normally are considered a hardware design issue, there are some areas that software must control. For example, in a system with multiple software-controlled motors, it may be possible for the software to self-induce EMI problems. DC motors have a larger current draw at startup, possibly three times the normal running current. If the software simultaneously starts multiple motors, the resulting current surge may disrupt the system electronics or even the processor itself. It may be necessary for the software to sequence motor startup and braking to prevent these problems. Software also may need to filter sensors that are susceptible to radiated interference. In general, it is better to provide such protection in hardware, but sometimes the nature of the sensor makes it impossible to filter the interference. This might happen if the sensor has to sense very low level signals or has to sense over a long cable. In cases like these, the software may need to filter some of the signal.

Interrupt protection. Sometimes the software needs to protect itself against spurious interrupts. One real-world case involves a motor with a reflective strip on the shaft (Figure 3.3). The strip is sensed with a reflective optical sensor to count and time motor revolutions. In some cases, the motor would stop with the reflective strip right on the edge of where the sensor could detect it. Vibration when the machine was running then would cause the sensor output to switch at a high rate, flooding the processor with interrupts. The problem is complicated by the fact that the resulting error code indicates that the processor is running out of time to complete its tasks—which it is, since it is spending enormous amounts of time in the interrupt routine. A similar case can occur when an operator not-quite closes a hood with a sensor on it or if a sensor is not securely mounted, leaving it susceptible to

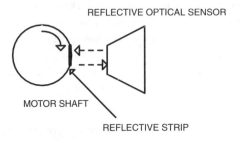

Figure 3.3
Unstable Interrupts from an Optical Sensor.

vibration. Any time you have moving objects detected by sensors, you run the risk of unstable output.

Bus width. Sometimes an 8-bit device is connected to a 16-, 32-, or 64-bit processor. As mentioned in Chapter 2, an example would be a 16550 UART connected to an ×86 processor. In most cases, the hardware is not designed to decode all the possible accesses to the device, so the software must not try to perform writes that are wider than the device. In some cases, this is no problem, as the unused bits will be discarded. However, if the hardware is designed so that the other bits write to a control register, you can get unexpected results. For instance, if a device is located at address 03F0 (hex), and the lower 8 bits are connected to a data register while the upper 8 bits are connected to a status register, a word write will change both registers. Another case where this can cause problems is on a read: If you read a 16-bit word from the 8-bit device, the unused bits usually will be undefined because those bits are floating. If you do not mask them off in software, the results are indeterminate.

Software Architecture

There are only so many ways to connect an erasable programmable read-only memory (EPROM) to a processor, but there are numerous ways to implement almost any software function. However, embedded software usually is built on only a few architectural frameworks, as described in the following paragraphs.

Single Polling Loop

In this method, a single piece of polling code loops continuously, checking for input from interrupt routines and external devices (such as a keypad) and executing whatever subroutines are necessary to implement the functionality. This method of coding assumes that all functions are available all the time. For example, it might check all the key switches all the time, even if some switches are not used in particular modes. This was the method that I used to implement the protocol converter described earlier. That design did not even have any interrupts—everything was done in the polling loop.

State Machine

The software is in one state at a time. Only those functions that pertain to the current state are monitored. I designed a burglar alarm this way once. The system had states that included armed, waiting for arm, triggered, and alarm on. This method has the advantage of compartmentalizing the functionality. You need not worry about some process getting confused as to what state everything is in, because every state has its own unique code. And you can change the code for a particular state without affecting the code for any other state. The disadvantages to this method, first, are that there usually is a lot of duplication. A keypad, for example, has to be monitored regardless of the state, so the key monitoring code (or at least a call to the subroutine) is duplicated in the code for every state that needs it. And, if a number of parallel processes can be in different states, you need a unique state for every combination, which makes the code grow exponentially. State machine architecture is best suited for designs that perform a single function.

Multiple State Machines/Polling Loop

In this variation on the state machine architecture, each process can have unique states. A polling loop goes to each process, which then branches to code for the particular state it is in. When done, the process code exits to the next process (or to the polling loop, which goes to the next process). The pool timer uses this method. The polling loop checks the motor state, to see if the motor should be on or off, then checks for timer rollover. What happens after that depends on whether the code is in powerfail, time set, or normal timekeeping mode (state). For example, if the code is in powerfail mode, the code for the other modes is not even executed.

Incremental State Machine

Each process executes a few steps of whatever operation it performs in whatever state it is in, then transfers control to the next process. Each process also keeps track of where in its internal sequence it is. The next time the process is executed, it takes up where it left off and executes a few more steps. This gives all processes the appearance of executing simultaneously, albeit slowly. I once had to find a bug in one of these that was written by someone else, and I am not fond of this method.

RTOS

Real-time operating systems warrant a dedicated overview in their own chapter. Basically, a RTOS allows the code to manage tasks by starting and stopping them based on priority or time. For example, the code to communicate with another system might be activated only if there are data to send.

The Development Language

It has been said that you cannot consider yourself a true embedded programmer unless you can code in assembly language. While this may be an exaggeration, it is true that many embedded systems have some code written in assembler. Part of this is because high-level languages (HLL) often assume that things like the stack are initialized. Since function calls in any language need the stack, which usually is in random access memory (RAM), you cannot do a function call to the function that initializes the stack or the RAM chip select logic. The other reason is speed. The best code optimizers still are not quite as good as a human at generating fast assembly code.

That said, high-level languages for embedded applications are becoming more sophisticated, more efficient, and better able to handle the unique hardware that embedded systems must control. The choice of development language, like every other engineering tool, is driven by trade-offs such as cost, ease of use, and utility.

I know that the following statement will make some software engineer throw this book across the room in disgust, but a simple project may be well served with just assembly language. Many small projects are not completed any faster by using a HLL. Some manufacturers (Microchip and Atmel, for example) provide free assemblers on their websites. But cost is not necessarily a factor in choosing assembler over a HLL. Sometimes the issue is speed. Sometimes assembly language is the only way to get everything to run fast enough. This is particularly true when using a DSP or other hardware-intensive processor. And, if most of the code is dedicated to flipping bits on I/O ports or loading timers, most of the design effort goes into calculating things like the timer values. The actual code may be fairly simple. Of course, the drawback to using assembler is that whoever has to maintain the code in the future must learn the architecture and instruction set of the processor.

I once used assembly for a small microcontroller project because I could download the assembler from the manufacturer's website and get the project going faster than I could get the paperwork through the purchasing department to order a C compiler.

As the complexity of a project grows and labor becomes a larger part of the software design, assembly language begins to look less attractive. The time spent debugging the code becomes more of a factor, and the ability to write code for more than one processor without learning a new language is important. However, some of this advantage often is lost on simple microcontrollers. The high-level languages for those devices often have limitations, such as lack of floating-point capability, that limits code portability. In addition, frequently special instructions are needed for controlling device-specific hardware (timers, I/O ports, etc.) that will not port to another processor.

I will not attempt to recommend the best HLL for your application. Every software engineer has a preference. What I will do is list some things that should be considered in the decision.

Processors Supported

C has become almost a universal language, available for nearly any micro-processor. PL/M and Pascal are nearly obsolete and not necessarily available for microcontrollers. As already mentioned, be aware that HLL versions for microcontrollers may not support the full language. Some things that are normal for a HLL on your desktop computer are just not possible with 128 bytes of RAM. Also, be sure that the compiler supports the version of the processor you use. An extreme example where this could be a problem is if you are developing code for a Pentium-class processor with a compiler that generates only 8086 code. You would be unable to take advantage of the added features in the Pentium CPU.

Emulator Support

Most current emulators, instead of displaying hex addresses, can display labels from the source code (called *source-level debug*). Instead of single-stepping through one machine instruction at a time, you can step through one C (or whatever) instruction at a time. You can even step through entire functions as if they were a single statement. You can set breakpoints the same way. This can reduce debug time *enormously*. However, the emulator software has to have a table of addresses versus labels. Be sure your language and emulator are compatible. This process is simplified somewhat by the emergence of stan-dard file formats for this data, such as UBROF. However, the emulator needs to handle whatever file format the compiler produces if you want to do source-level debug. Ideally, you want the debugger to display the data and in the correct format, recognizing 8-bit character variables, integers, and other data types.

Code/Storage Size

Some compilers are extremely inefficient in their use of PROM and RAM resources. Be sure the compiler will not require enormous increases in hard-ware cost to support its free-spending ways. The same goes for speed: Some compilers just cannot seem to generate fast code.

Optimization

Optimizing compilers attempt to produce the most efficient code and can make the difference between an application that works and one that does not. An optimizing compiler works by eliminating unnecessary machine code. For example, if the code tests a variable to see if a bit is set, then tests the variable again to check a different bit, a nonoptimizing compiler might read the variable twice. An optimizing compiler might read the variable once, store it in a register, and knowing that the variable has not been changed, use the value in the register when the second check is done. But beware; this can cause problems. If the variable actually is a hardware register, a change between the first and second checks would go undetected in the optimized version. For this reason, most optimizing compilers allow optimization to be turned off for sections of code, and they allow you to define memory-mapped hardware that is treated differently than ordinary RAM. Some compilers, especially for microcontrollers, permit you to optimize for speed or size, but there sometimes is a catch: You can optimize only the entire program. No compiler directives let you turn optimization on and off. To offset that, microcontroller compilers typically have a way to define hardware-specific addresses, such as I/O ports, forcing the compiler not to optimize when accessing these locations.

If your application requires floating-point calculations, be sure the floating point libraries are small enough to fit the available space and that they run fast enough. If you have a hardware floating point processor, of course, be sure the compiler can take advantage of it.

Assembly Support

Many applications still require assembly language for things like initialization, interrupts, or specialized (fast) I/O. The development compiler should make it fairly easy to include assembler files (possibly as inline code) into the software.

Another assembly-language related issue occurs when using microcontrollers. Due to the limited stack size, high-level languages typically do not pass parameters to assembly functions using the stack but instead use an area in RAM. This usually is an overlay area shared with other functions, because the information no longer is needed when the function is finished. When writing a function in the HLL, the compiler takes care of getting dynamic variables out of the RAM overlay area, initializing the variables within the function, and restoring everything when the function is finished. If you have to link assembler functions into HLL code, you must take care of all these details. So be sure the HLL compiler has a well-defined interface to the assembler and a well-documented procedure for defining the functions in the HLL and then writing the functions in assembler. You can waste a lot of time trying to get this to work if the vendor does not document things well.

Another problem with microcontroller-based compilers is that the compiler tends to generate code that is generic—the code produced for a do/while loop may not take advantage of instructions that can speed up the code considerably if you are ending when a variable reaches 0 or something similar.

An example involves a design I did using a Microchip PIC device. The processor functions as a smart sensor, communicating with an external PLD that collected time-based data. To make the PLD interface work, the processor needs one loop to run very fast. The C compiler I was using simply would not generate any kind of loop that terminated with a DECFSZ instruction, which decrements a memory location, stores the result, and branches if the result is 0. In addition, a feature of the PIC that involves a shift-and-test operation was implemented inefficiently. To make matters worse, the assembly language interface for that compiler was poorly documented, making the task of linking an assembly module very difficult.

The original code, prototyped in assembly, took less than a day to write, debug, and verify. Attempting to write a C version, then to merge the original assembly version with the final C program took three days and numerous e-mail exchanges with the technical support group at the vendor that supplied the compiler.

While the language is the tool that actually implements a design, other tools document it. Flowcharting tools simplify flowchart generation, with linkages between blocks "rubberbanding" as the blocks are moved around on the page. Some tools offer a complete development environment, from flowchart to finished code.

Debugging Tools

When selecting tools, give some thought to how the design will be debugged. If you are developing an embedded system based on the PC architecture, you probably will get some kind of debugger with the compiler. However, on a smaller design, there are other considerations. The most severe restrictions occur when you are dealing with microcontrollers. For example, some debugger software for microcontrollers requires use of the on-chip serial port for communication with a host PC. If your design uses the serial port to communicate with other processors in the system, then you have to find another debugger solution.

Even some emulators require system resources (such as serial ports) that you may need for your design. Check into this carefully before you choose your development environment. Make sure you do not paint yourself into a corner.

Debugging systems on very small microcontrollers presents special challenges. The best solution is an emulator. However, sometimes an emulator is not available for cost reasons. I have worked on microcontroller-based designs where an emulator simply would not fit in the space available or the controller board was on a moving robotic arm and attaching an emulator was out of the question.

In cases like this, you have to make special concessions. Most small microcontrollers have a *simulator* available. This is software that runs on a PC and emulates the microcontroller. A simulator allows you to develop code and test it right at your desk.

Of course, simulators have one serious drawback—they cannot predict the real world. A simulator can only simulate with the inputs you provide, and if you cannot predict all the real world timings, neither can the simulator software. On the other hand, if your microcontroller design is doing something that is predictable and not dependent on real world timing, a simulator may be the only debug tool you need.

The real environment, with moving motors and clicking solenoids is where most embedded designs end up, and this is where the limitations of a simulator are really revealed. If a simulator is not adequate and if you cannot get or cannot use an emulator, then you have to turn to other debugging techniques, such as those described in Chapter 5.

Microprocessor Hardware

An embedded system has unique hardware constraints that must be accommodated by the software.

The Stack

The stack, common to nearly all microprocessors, is a place where the software can temporarily store values until they are needed. When a subroutine is called, the processor saves the return address on the stack. The stack is a last in, first out (LIFO) buffer—the last value placed "on" the stack is the first value removed. Like those spring-loaded stacks of plates in a restaurant buffet line.

Most (but not all) microprocessors have PUSH and POP instructions, or some equivalent, that can add values to or remove values from the stack. This allows the programmer to save registers during an interrupt and restore them later. The stack in your PC can hold thousands of bytes of information. The stack in an embedded design is not always as flexible:

- First, the stack in some microcontrollers is limited. The PIC17C42, for example, has a 16-level stack and no PUSH or POP instructions. The stack is used for return addresses only.
- In processors that have a hardwired stack (implemented as fixed registers in the microprocessor IC), you cannot have several levels of subroutines; the stack will overflow. As already mentioned, you cannot pass parameters on the stack (at least, not very many) for the same reason.
- In processors that *do not* have a stack limited by hardware, there is still a limitation, the size of the system RAM. Make sure that the code cannot make the stack grow into the area where variables are stored. This problem can be hard to find.

Getting around a limited stack sometimes requires programming finesse that bends the normal rules a bit. The usual practice for a subroutine is to save all the registers on the stack; some high-level languages do this automatically. This can be impractical if there are several subroutines.

The simplest workaround is not to save anything. The polling loop just knows which registers are used by the subroutine and assumes they will be changed. Values can be returned in registers as well. The pool timer code has subroutines that do not save registers.

In cases where registers must be saved, they can be stored in RAM as variables. Each subroutine has a block of RAM set aside for storing registers. Each register that must be saved is stored in a unique RAM location on entry to the subroutine and retrieved on exit. Of course, this method prevents the subroutine from being reentrant.

Some processors make provision for a context switch. The Analog Devices ADSP2101 family, for example, has two complete register sets. Either may be selected by a single instruction. The 8051 has four register banks. Two bits in an internal register control which register bank is in use. In processors like these, each subroutine can have a unique register set. But there is a drawback as well; the Analog Devices parts have only one extra register set, so only one subroutine can be handled without saving registers into RAM. The 8051 has four register sets, which limits the code to four unique environments.

Although it is easiest to take advantage of programming techniques like this in assembler, many compilers can also produce this sort of efficient code.

Chip Select

Some processors have internal chip selects that can be programmed. If this hardware is used, it must be initialized first. As an example, let us look at the

80188 again. This part has a signal, -UCS, which is a chip select to upper memory, intended to select an EPROM. After a reset, -UCS is programmed to access the upper 1 K of memory from FFC00 to FFFFF. If the actual EPROM is 32 K × 8, the software at the reset vector location (FFFF0) must initialize -UCS to select the full 32 K before jumping to any location below the upper 1 K. Otherwise, the EPROM will not be selected and the processor will go off into the weeds. Staying with the 80188 example, the signal -LCS is intended to select RAM from location 00000 up. Reset turns this signal off, and it must be programmed before it will be active.

The implication of this is that, in a system using -LCS and -UCS, *no* subroutine calls can be made until both signals are initialized. If the code that initializes -LCS, for example, is a subroutine call, an attempt is made to push the return address onto the stack before the RAM (where the stack is located) can be accessed. The return at the end of the subroutine ends up with a garbage address and the processor goes off into the weeds. The same arguments apply to any programmable chip select that affects PROM or RAM.

RAM

Most high-level languages allow local variables, which are completely local within a subroutine or function. Other processes know nothing of those variables and have no access to them. In a PC, memory management hardware can even tell the operating system if an ill-behaved program steps outside its predetermined boundaries. In an embedded system, local variables often are an illusion. As already pointed out, embedded systems have constraints on RAM size. A variable is just one or more memory locations. All variables are stored in the same RAM space, often in the same RAM integrated circuit (IC). Each variable has a unique location, but nothing will prevent a berserk piece of code or an incorrect pointer from changing the wrong location or even writing all through memory and trashing everything. Assembler programmers are familiar with this concept, since variables in assembly language (especially with simple assemblers) are often global. Unless there is hardware memory management, all variables are potentially global to incorrect code.

When using languages that support reentrant code and genuine local variables, saved on the stack, be sure there is enough stack space.

Initialize all variables. In some nonembedded applications, variables such as tables can be initialized when the code is loaded. This is not true in an embedded system. Initialize everything.

One last note about RAM and PROM: HLL compilers for embedded use often require that you tell the compiler (or actually the linker) where the RAM and PROM are located, what size they are, and where the stack is to be (if used). This information usually is used by the linker, a program that links together various code modules and produces a single output file. Some compiler/linkers are capable of calculating the required stack size (or at least the worst-case stack size) and automatically setting the stack pointer appropriately.

I/O Ports

Programming hardware I/O ports for a microcontroller varies depending on the particular part used. Some microcontrollers, such as the PIC17Cxx, have control registers that control the direction of each port bit. Others, such as the 8051, make every port bit an input at reset and the software makes the port bit an output by writing to it. I/O ports for LSI I/O ICs usually have a direction register.

Switch and Other Contact Closures

A switch closure such as relay contacts or a push-button switch will "bounce." Always. A switch "bounces" when opening or closing and takes the form of repeated openings and closings of the contacts. If a switch is not debounced in hardware, it usually will need to be debounced in software. Most mechanical switches finish bouncing in 10–20 milliseconds (ms). The logic to debounce a switch is as follows:

Switch closure detected:

Wait 30 ms.

Check switch again. If open, it was bounce on opening. If still closed, it was a valid switch closure.

Dangerous Independence

An earlier example showed a protocol converter with parallel, independent processes. Independent processes are common in embedded systems, but it is also important to know when the processes should *not* be independent. A real world example will illustrate this: I worked on a motion control subsystem, developed by a contractor, that feeds items into a moving transport system (Figure 3.4). The system has two requirements that are relevant here. The first is to meet a particular throughput in items per minute. The second requirement is to maintain a minimum spacing between any two successive items. Two independent processes run these two functions. The first process controls a terminal motor speed to maintain throughput, and the second

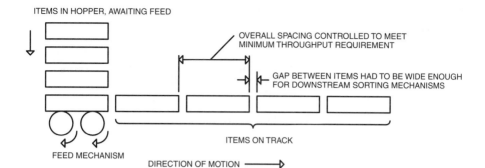

ITEMS IN HOPPER, AWAITING FEED

OVERALL SPACING CONTROLLED TO MEET
MINIMUM THROUGHPUT REQUIREMENT

GAP BETWEEN ITEMS HAD TO BE WIDE ENOUGH
FOR DOWNSTREAM SORTING MECHANISMS

ITEMS ON TRACK

FEED MECHANISM

DIRECTION OF MOTION

Figure 3.4
Conveyor Belt System.

process performs instantaneous adjustments to the motor speed to maintain minimum spacing between individual items. These corrections are small or large, depending on the predicted spacing error. The corrections reduce throughput but are expected to occur infrequently. After a correction, the motor ramped back up to the terminal speed.

The prototypes of the system worked well, but production units could not meet the throughput requirement. It turned out that the spacing process was performing a few corrections to fix spacing and this reduced throughput. The throughput process, not knowing about corrections, saw that throughput was too low and raised the terminal speed. This required even more corrections, which lowered throughput further. The result was that the throughput process walked the speed up to the maximum value, while the spacing process corrected more and more items, until it was finally correcting (or overcorrecting) nearly every item. The reason this showed up only in production was that the mechanical adjustments in the prototype allowed the system to "balance" and run without problems.

The solution for this problem was fairly simple: If too many corrections were performed, terminal speed was lowered instead of raised. The amount of speed reduction was weighted by the degree and number of corrections that had occurred. If a problem with the mechanical components or the condition of the items processed caused a lot of corrections, throughput went down until things stabilized again. The processes no longer were independent, but that was necessary to prevent positive feedback.

By now, some readers are pointing out that these two software processes were not truly independent, and that is true. They were coupled, but only through the mechanical characteristics of the system, and that kind of problem can be very difficult to isolate.

Software Specifications

The software specifications tend to be the one document never produced or produced at the end of the project and in a hurry. This is for several reasons; first, there is no "customer" for the software specs like there is for the hardware specifications. The hardware specs are needed in some form so the software can be written, but there often is no corresponding need user for equivalent software information.

The software often is the last part of the project started, and it often starts late. The software engineers may be late because a previous project ran late, and the rest of the project may have had changes right up to the very end. Since functional changes often are implemented in software, it is difficult to finish the code while the project is still in a state of flux.

Finally, tools are available that document a software design from flowchart through release. If you can describe the actual software as you go, why not just use that for documentation?

Since there often is no downstream "user" for the software specification, what are the reasons to create one? Who will read it? Choose one of the following reasons:

- The software specifications serve as an overview of the code for the software engineer who has to maintain it. This is important if the code is maintained by someone other than the person who wrote it. After a year or so, it will also become important to the person who wrote the code.
- On a multiprogrammer project, it helps coordinate the effort.
- It is a useful document for design reviews and for checking that the software functionality really matches up with the hardware capability.
- It can uncover oversights and conflicts in the preceding documents (hardware specs, requirements, etc.).
- It can help clear up confusion regarding actual functionality, such as how the operator interface works.

Software Specifications Outline

The following is a generic outline for software specifications.

Overview. A brief description of what the specifications cover.

Related documents. ANSI, IEEE specs. May also include a reference to the product requirements document.

Description. A brief description of the software and what it does. Typical sections include

Operator interface. A detailed description of operator interaction with the system. How a keypad is used, what screens are displayed, what keys are locked out in which modes, etc.

Interfaces to other systems. Includes a description of protocols used for communication. Proprietary protocols should be spelled out in detail, with opcodes defined, checksum/CRC methods defined, sizes of data packets specified, and the like. Spell out things like how word-wide data is transferred over byte-wide interfaces (LSB first or MSB first?). Standard protocols (such as TCP/IP) simply can be referenced.

Hardware controlled. This may appear redundant, since the hardware specifications also cover it. However, the software specifications should detail how the hardware will be controlled. Specific algorithms should be specified. If the algorithm is very complex, it may be included in an appendix and referenced here. Assumptions about the hardware should be included, such as the following examples:

Does the motor controller have hardware protection against shoot-through when the motor changes direction or must this be done in software?

Is a software interlock required to prevent two relays from overlapping?

How much time does the software allow for the relay to open before checking status?

To what state is the hardware assumed to be initialized?

How long is the software-generated reset pulse held?

Will the software need to insure a write recovery time for any hardware devices?

Interrupt/task priorities. Unique requirements for priorities of interrupts or tasks should be documented. Nested interrupt requirements should be specified.

Interrupt usage. How interrupts are used, which are used for what, which ones are edge versus level sensitive, and so forth.

Memory usage. How the memory will be organized, how much is required for data buffers, how much is needed for storing acquired samples, and so on.

Tools. What software tools will be used for development.

Special requirements. For example, if the software will perform a lot of floating-point calculations, this should be spelled out. If there is no hardware FPU, then you will want to prove that the system throughput will still be adequate.

Appendix. The appendix should include any calculations that go into the software design, such as the following:

Maximum interrupt latency for critical interrupts. How long interrupts can be turned off.

Service time calculations. To verify that processing does not fall behind. Things like processing time for serial data to be sure a byte is not missed.

Interface speed. Calculations to verify that, say, the Ethernet can keep up with the data.

Data movement. If you move a lot of data around, calculations to verify that the PCI bus, for instance, has sufficient bandwidth to handle all the data.

In Chapter 4, we look at interrupt hardware and software, which is typically what makes an embedded system into a real-time system.

Interrupts in Embedded Systems   4

I have deliberately left discussion of interrupts out of the preceding chapters, treating interrupt code as a "black box" that produces certain outputs. Part of the reason for the brevity of Chapter 3 is that the real-time aspect of an embedded system often depends on interrupts, and such software is covered here. Interrupts in a real-time system require a tight relationship between software and hardware.

Interrupt Basics

An interrupt is an input to a microprocessor that temporarily redirects the program flow. An interrupt can notify the processor when an analog-to-digital (A/D) converter has new data, when a timer rolls over, when a direct memory access (DMA) transfer is complete, when another processor wants to communicate, or when almost any asynchronous event happens.

The interrupt hardware is initialized and programmed by the system software. When an interrupt is acknowledged, that process is performed by hardware internal to the processor and the interrupt controller IC (if any). When an interrupt occurs, the on-chip hardware performs the following functions:

- Saves the program counter (the address the processor was executing when the interrupt occurred) on the stack. Some processors save other information as well, such as register contents.
- Executes an interrupt acknowledge cycle to get a vector from the interrupting peripheral, depending on the processor and the specific type of interrupt.
- Branches to a predetermined address.

The destination address is the *interrupt service routine* (ISR, or sometimes ISP for interrupt service process). The ISR performs whatever functions are

113

required and then returns. When the return code is executed, the processor performs the following tasks:

- Retrieves the return address and any other saved information from the stack.
- Resumes execution at the return address.

The return address, in nearly all cases, is the address that would have been executed next if the interrupt had not occurred. If the hardware and software engineers do everything right, the code that was interrupted will not even know that an interrupt occurred.

Interrupt input to microprocessors comes in various flavors. Some processors have dedicated interrupt input pins that send the processor to a specific address. Other processors have only one interrupt pin, and the interrupting device must supply an *interrupt vector* that tells the processor where the ISR is located. Some processors have both kinds of input.

Interrupt Vectors

All processors require an interrupt vector when an interrupt is acknowledged. The interrupt vector tells the processor where to go to service the interrupt. On some processors, the vector can be an actual instruction that is executed just as if it occurred in ordinary code. Other processors expect a number that is translated by the processor into an address.

Figure 4.1 shows the three methods of generating a vector to the processor: from an external interrupt controller, from an internal interrupt controller, or from the peripheral itself. On some processors that use an internal controller, the internal hardware may not produce a separate interrupt cycle with a true vector, but the effect is the same.

Edge- and Level-Sensitive Interrupts

An interrupt input can be edge or level sensitive. A level-sensitive interrupt is recognized by the processor whenever the interrupt pin is in the active state. An edge-sensitive interrupt means that the processor responds to a rising or falling edge on the interrupt pin. Some processors and interrupt controllers have interrupt inputs that can be programmed as either level or edge sensitive. Edge- and level-sensitive interrupts are addressed in more detail later in the chapter.

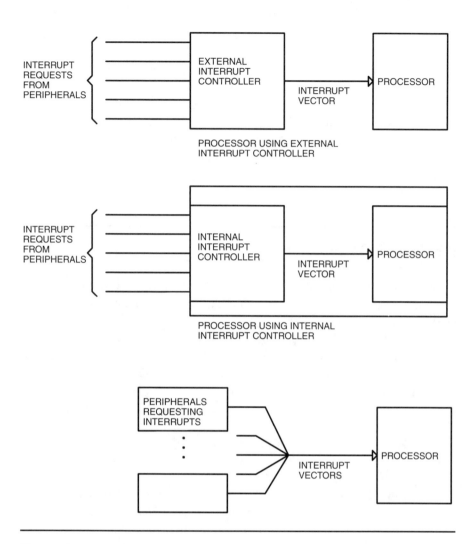

Figure 4.1
Interrupt Vector Generation.

Interrupt Priority

Interrupts usually have a priority. The priority determines when an interrupt is serviced. A higher-priority interrupt takes precedence over a lower-priority one if both are asserted at the same time. Some processors permit *nested* interrupts. When enabled by software, nested interrupts allow an ISR itself to be interrupted by a higher-priority device. Interrupts from lower-priority devices are ignored until the higher-priority ISR is completed.

Different types of processors have different priority schemes. The 68000 family parts, for example, allow a peripheral requesting an interrupt to assert its own priority, and it is up to the hardware engineer to make sure there are no conflicts. The Intel 8259 interrupt controller has several interrupt input pins, with programmable priority. The priority of interrupts inside an embedded processor sometimes is fixed, sometimes programmable.

Interrupt Hardware

Hardware to implement interrupts varies with the processor and the peripheral doing the interrupting. We will look at the simplest cases first and work up.

The simplest interrupt is a single pin on the microprocessor. The 80188, for example, has four interrupt inputs: INT0, INT1, INT2, and INT3. (There is a fifth, NMI, which we discuss later.) These interrupts may be programmed to be either level or edge sensitive. Each pin, when activated, causes the processor to vector to a specific address, as shown in Table 4.1.

The code at the interrupt address is usually a jump to the actual ISR somewhere else. The vector, as mentioned earlier, is provided by an interrupt controller inside the 80188.

The software may enable any, all, or none of these four interrupts; and it determines if the interrupts are level or edge sensitive. In some processors, such as the PIC 17C4x, interrupt pins can be used for functions other than interrupts, and this is under software control as well. In general, most proces-

Table 4.1
80188 Interrupt Vector Addresses.

Interrupt	Vector Address
INT0	00030h
INT1	00034h
INT2	00038h
INT3	0003Ch

sors have the ability not only to enable and disable specific interrupts but to disable all interrupts at the same time.

The second type of interrupt is generated by internal peripherals. The 80188 internal peripherals that can generate interrupts include two DMA controllers and three timers. Other processors and other versions of the 80188 have different internal peripherals, such as universal asynchronous receiver/transmitters (UARTs).

Internal interrupts work much the same way as external interrupts. Some event, such as a timer rollover, occurs and an interrupt is generated to the processor. Like the external interrupts, these usually have a predetermined interrupt vector address. The software must enable the peripheral device and enable interrupts from the device. Like interrupts from the external pins, these interrupts are handled and the corresponding vectors produced, by the internal interrupt controller.

Sometimes internal interrupts are shared. That is, multiple devices may share a single interrupt source and vector. The timers in the 80188 work this way—all three timers use the same interrupt. The ISR must read the timer status bits to determine which timer (or timers) generated the interrupt. Similarly, the PIC 17C42 has several internal peripherals that can generate interrupts, but only three interrupt vectors, so several peripherals must share an (internal) interrupt signal and vector. When a peripheral interrupt occurs on the shared line, the ISR must poll the interrupt status bits to see which peripherals are requesting service. The PIC 16C6x series has only one interrupt vector—all interrupts require polling to determine the source, unless only one source is enabled.

The next level of interrupt complexity involves a vector provided by the peripheral. In this scheme, the peripheral interrupts the processor, and the processor acknowledges the interrupt. When the acknowledge occurs, the peripheral places a vector value on the data bus. In some early processors, such as the 8085 and Z80, the vector was actually an instruction. These processors include several 1 byte instructions that force the processor to a specific address in low memory, and the interrupt controller typically would provide one of those instructions in response to an interrupt acknowledge. Other processors expect a number, which is used as a pointer into a table, usually in low memory, that contains the jump instructions to the ISRs. (The Z80 actually supports both schemes.) The 8086 family, for example, reserves the first 1024 bytes of code space, from 00000 to 003FF, for the interrupt table.

Using external vectors on processors that have that capability is the most flexible interrupt scheme, but it also requires more hardware than other methods. Each peripheral in this scheme must include the hardware to recognize interrupt cycles and provide the correct vector.

Interrupt Bus Cycles

When using an external vector, most processors perform an interrupt acknowledge cycle that is similar to other bus cycles, but with different control or status signals. The 8086, for example, performs a bus read cycle but with the interrupt acknowledge (-INTA) signal replacing -RD and a different status indication. The 68000 asserts -DS normally, and only the status bits indicate that the bus cycle is an interrupt acknowledge.

Some designs use an interrupt IC that has several discrete interrupt inputs but interfaces to the single processor interrupt line. These ICs produce the interrupt request to the processor and, when the interrupt is acknowledged, return the interrupt vector corresponding to the highest-priority interrupt pin activated. The 8259, typical of interrupt controller ICs, has eight interrupt inputs and is designed to interface to Intel processors.

Daisy-Chained Interrupts

Some designs use daisy-chained priority interrupts, which are illustrated in Figure 4.2. The interrupt input in this scheme, usually open collector, is driven by all the interrupting peripherals. When any peripheral needs to interrupt the processor, it activates the interrupt line. Each peripheral also

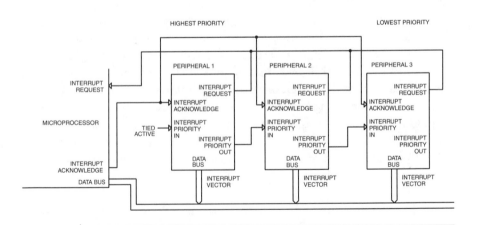

Figure 4.2
Daisy-Chained Priority Interrupts.

has a priority in and a priority out. The priority in is connected to the next highest peripheral in the daisy chain or tied active for the highest-priority device. The priority out is connected to the next lowest peripheral. If any particular peripheral is not requesting an interrupt, its priority out follows priority in. If the peripheral is requesting an interrupt, its priority out is blocked, which also forces all lower priority out pins to be blocked. This prevents lower-priority peripherals from generating an interrupt vector. When the processor generates an acknowledge, the highest-priority peripheral that is requesting an interrupt returns the vector.

The advantage of daisy-chained interrupts is that fewer interrupt lines are needed—one instead of one per peripheral. The disadvantage is the priority structure. In this method, the first device has the first chance at acknowledging the interrupt, even if it asserted the interrupt after a lower-priority device. If several higher-priority devices (that is, higher in the acknowledge chain) assert interrupts, it may be some time before the interrupt from a lower-priority device gets serviced. The priority of each device in a daisy-chained scheme is fixed by the peripheral's position in the chain—the software cannot change it.

Other Types of Interrupts

The 68000 family of processors can support a fairly sophisticated interrupt scheme. The processor has three interrupt input signals that are encoded into seven priority levels (the eighth level occurs when all the inputs are inactive). Any device can request an interrupt by driving these inputs with its priority level. A priority circuit inside the processor checks the input priority against the current processor priority. If the input priority is higher, the interrupt is acknowledged. The interrupt vector may come from the interrupting device or may be generated internally. If the interrupt request input is of a lower priority than the current processor priority, the interrupt is ignored.

The interrupt inputs are often driven by open-collector devices so that multiple peripherals can share them. Multiple peripherals can share the same priority level, but each peripheral can provide a different vector to the CPU.

In addition to ordinary interrupts, many processors also have a nonmaskable interrupt (NMI) input. As its name implies, NMI is not maskable (cannot be ignored) by the software and always will be serviced by the processor even if interrupts are off. It normally is used for things such as pending power shutdown, memory parity error, or some fatal error in the system. Many

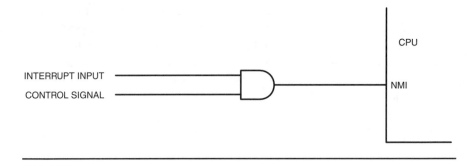

Figure 4.3
Gating Used to Enable and Disable NMI.

embedded designs do not use NMI but terminate it in the inactive state so it can never occur. NMI *can* be used just like any other interrupt, but remember that it cannot be ignored; when NMI occurs, the processor always responds. If you are using NMI as a general-purpose interrupt, be sure that you know when it can occur or make provision to disable it with external hardware. Make sure the NMI cannot occur before the interrupt vector and stack are set up.

Figure 4.3 shows how the NMI can gated with external hardware using a two-input AND gate. One input of the gate goes to the interrupt signal. The other input goes to a control signal, which is a port bit or register bit that can be written by the CPU. The output of the gate goes to the NMI input on the CPU. To enable the interrupt, the CPU sets the control bit to a 1. This allows the interrupt input to drive the NMI line. To disable the interrupt, the CPU sets the control line to a 0. This example assumes a high true interrupt source and a high true NMI; obviously the logic would be different if the input or NMI was low true.

Using Interrupt Hardware

This section describes some guidelines for interfacing to the interrupt hardware.

Level-Sensitive Interrupts

When using level-sensitive interrupts, remember that the processor will see an interrupt when *and only when* the level-sensitive interrupt pin is in the active state. This can be an advantage or a disadvantage. If the interrupt gets stuck

Embedded Microprocessor Systems

in the active state, the processor will service the interrupt, exit the ISR, and immediately reenter the ISR. On most processors, nothing else will get done, as the processor loops continuously in the interrupt code. When using level-sensitive interrupts, make sure they cannot get stuck. If the interrupt comes from an external device or system, be sure that turning off the power to that device will not leave the interrupt in the active state. Devices that generate level-sensitive interrupts usually need some mechanism that allows the processor to clear the interrupt request before exiting the ISR. This may happen automatically, such as when the processor reads a byte from the peripheral. If clearing the interrupt does not happen automatically when the ISR is executed, you may need to write to a command register in the peripheral. A discrete hardware implementation may need a flip-flop that generates the interrupt and can be reset by the firmware.

The reverse of a stuck interrupt also can occur with a level-sensitive interrupt. If the interrupt is asserted and removed before the processor services it, the interrupt (usually) will never be recognized. This can occur if the interrupt is not latched and the processor has interrupts disabled or if the processor is busy servicing a higher-priority interrupt that takes longer to handle than the active time of the missed interrupt.

Level-sensitive interrupts can be useful if multiple devices share the interrupt. The devices can each assert the interrupt when necessary. If two devices assert the interrupt at the same time, the processor will service the first one, exit the ISR, reenter the ISR, and service the second. Note that, in processors with a single interrupt input and an external controller or that use daisy-chained interrupts, the single interrupt input is level sensitive.

Edge-Sensitive Interrupts

Edge-sensitive interrupts are ideal for counting events. The processor accepts the interrupt only on the edge. Some processors have a requirement that the interrupt go to the active state and stay in that state until serviced. For these devices, pulsing the interrupt to the *inactive* state generates the interrupt. For example, the 80188 interrupt inputs, when programmed to be edge sensitive, must go high and remain high until serviced by the processor. The interrupting device can leave the line *high* and pulse it *low* to generate an interrupt. The line can be left high between interrupts. If this technique is used, be sure that the inactive pulse width meets the minimum requirement for the processor or interrupt controller IC. This time usually is measured in clock cycles.

Figure 4.4 illustrates the difference between edge- and level-sensitive interrupts.

Edge-sensitive interrupts are ideal for applications where the peripheral needs to interrupt the processor without waiting to see if the interrupt

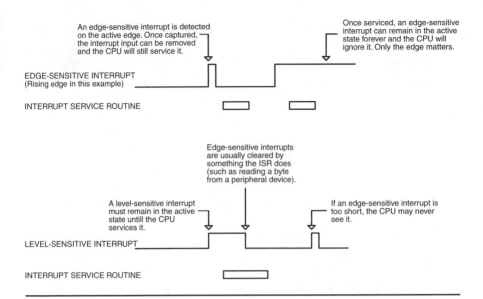

Figure 4.4
Edge- versus Level-Sensitive Interrupts.

actually is acknowledged. In addition, edge-sensitive interrupts have no problem with continual interrupts if they get stuck, but they have the opposite problem. If the interrupt gets stuck in the active state, the processor does not continuously service it but instead ignores subsequent interrupts. If two devices share an edge-sensitive interrupt and one device generates an interrupt request followed by the second device, the second interrupt usually will be missed since the processor (or interrupt controller) saw only one edge. Edge-sensitive interrupts rarely are shared for this reason. Figure 4.5 illustrates this condition.

If edge-sensitive interrupts must be shared, there are ways around the missed interrupt problem. The simplest method is to have a status buffer that can be read by the processor to see which devices are requesting an interrupt. Each peripheral must set its bit in the status buffer when it requests an interrupt and leave the bit set until the interrupt is serviced. The software services the first interrupt, enables the interrupt (if disabled), then checks the register. If more interrupt requests are pending, the processor services the second interrupt *before* exiting the ISR.

Another method to handle this problem is to arbitrate the interrupt input so that no device can request an interrupt while the line is active. Each interrupting device must hold the line active by using a resettable flip-flop or something similar until the processor clears it. The software does not clear the interrupt line until it has serviced the interrupt and is ready to accept another.

INTERRUPT 1

INTERRUPT 2

SHARED INTERRUPT
TO MICROPROCESSOR

ISR FOR INTERRUPT 1

ISR FOR INTERRUPT 2

ISR 2 OCCURS WHILE ISR 1 IS STILL HIGH.
MICROPROCESSOR SEES ONLY ONE EDGE,
CAUSED BY ISR 1, AND INTERRUPT 2 IS
NEVER SERVICED.

Figure 4.5
Shared Edge-Sensitive Interrupts.

Most edge-sensitive interrupt circuits permit the internal interrupt to be cleared before another actually is enabled, which keeps the ISR from being a reentrant.

A single device connected to an edge-sensitive interrupt can have the same missed interrupt problem as multiple devices if it generates interrupts at a rate faster than the processor can service them.

Many processors, such as the 80188, have timer inputs that can be used as interrupts. The timer is programmed to count external edges on the input pin and generate an interrupt on rollover. Then the timer is loaded with a count that is one less than the rollover value. The first edge causes a rollover and generates an interrupt. The ISR must reload the counter for the next interrupt. If the timer function is not needed, the input can function as an interrupt. Even if the extra interrupt is not needed because you ran out of interrupt inputs, using a timer input as an interrupt can be useful in some applications. If it is essential to know if an interrupt is missed, this can be determined by looking at the timer in the ISR. If no interrupts are missed, the timer will have rolled over to 0. If one or more interrupts were missed, the timer will keep incrementing them and the value will be greater than 0. If you have a use for this technique, remember that some timers can be programmed either to keep incrementing after a rollover or to roll over and stop. Be sure you program the timer to keep going after a rollover. Figure 4.6 shows a timing diagram of the use of a timer as an interrupt source.

Externally Vectored Interrupts

Using externally vectored interrupts, where the interrupting device generates the interrupt vector, is fairly straightforward. When using an interrupt controller IC, choose one that is compatible with the processor family and

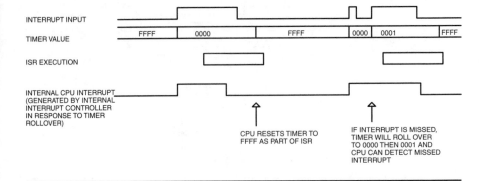

Figure 4.6
Using Timer Input as an Interrupt.

program it accordingly. A version of the interrupt controller IC must be selected that matches the speed of the processor. Be sure that you know the quirks of the part before using it. Some interrupt controller ICs, for example, can have *all* interrupt inputs be edge or level sensitive, but the individual inputs are not programmable.

In cases where there is no interrupt controller but the peripheral itself generates the vector, the design is a bit more complicated. First, the peripheral device must generate a vector that is compatible with the processor. In many peripheral ICs, this is programmable, but make sure that there are no conflicts with other devices that might use the same vector. Also, the timing must be compatible with the processor—setup time, hold time, and so on.

When interfacing a peripheral that generates vectored interrupts, be sure that other logic does not mistake an interrupt acknowledge cycle for a normal bus cycle. For example, if an address decode generates a read strobe to an input buffer, be sure it decodes the processor status lines and does not go active when an interrupt cycle is executed.

If daisy-chained interrupts are used and peripherals are not all from the same family, make sure that the priority in/out signals and timing delays are compatible.

When interfacing a peripheral that expects an INTA signal (such as 8086 family parts) to a 68000 or similar part that does not generate an INTA, the INTA signal must be synthesized by decoding the -DS and the processor status bits. Be sure that the decoding logic never mistakes a normal processor cycle for an interrupt cycle.

Daisy-chained interrupts have one disadvantage—they cannot be individually masked at the processor. If the peripheral logic has no provision to mask the interrupt, there is no way for the processor to ignore that interrupt when performing a high-priority task that cannot be interrupted. Be sure this is not a problem with the system design. If it is, some mechanism must be added to allow the processor to individually mask the peripheral interrupts.

The 68000 family parts, as mentioned earlier, have encoded interrupt lines. Most peripheral ICs have only one interrupt output. If you are interfacing just one of these to a 68000 family processor or other processor with encoded interrupt lines, you can wire-OR the lines to get the priority that you want. However, if there is more than one peripheral and you need more than one priority level, you have to drive the lines individually, through buffers.

Some peripheral ICs (such as the Z853x parts) require that INTA or other signals be synchronized to a clock. Be sure you work out this timing, and add wait states if necessary.

Interrupt Software

More than any other single thing, the handling of interrupts is probably what sets real-time embedded software apart from other microprocessor-based software.

If interrupts are used, the software must initialize the hardware. This means enabling the interrupts, loading the interrupt vector table into RAM (if required), programming vectors (in peripherals with that feature), and selecting any other relevant parameters (such as the edge/level mode). It is important that interrupts not actually be enabled until everything else is set up. If an interrupt occurs before the vector table is loaded into RAM, the results will not be good.

Interrupt Service Mechanics

When an interrupt occurs, the ISR is executed. All ISRs must perform three actions:

Service the hardware that generated the interrupt.

Enable the system to accept further interrupts.

Return control to the point at which the interrupt occurred.

The only exceptions to this are terminal interrupts, such as NMI, that can be used to signal the processor to stop due to an error.

Servicing the hardware that generated the interrupt means clearing the interrupt request, if necessary, and processing whatever caused the request. The pool timer system has one interrupt, a timer interrupt that occurs 250 times per second. Servicing that interrupt involves the following sequence (the full pseudocode is listed in Appendix A):

- Increment the $\frac{1}{250}$ second time counter.
- If the display should blink, do it at $\frac{1}{2}$ second intervals.
- At the 1 second rollover, decrement the current time.
- If the current time rolls over to 0:0, set a flag to tell the polling loop about it and load the next (off or on) time.
- Write the next display digit to the display and enable the proper digit. (This scans the display.)
- If keys are pressed on the keypad, debounce them and set the appropriate flag for the polling loop.

On returning from the ISR, enabling the system to accept further interrupts may be as simple as executing an interrupt return instruction, similar to the ordinary subroutine return but that also reenables interrupts. Other processors require a separate interrupt enable (reenable) instruction prior to an ordinary subroutine return instruction.

Some interrupt controllers, both those inside and outside the processor IC, need to be told when interrupt processing is complete, by writing a value to some address. In this case, just returning from the ISR is insufficient.

Returning to the code that was interrupted also means restoring the state of the machine. Any registers that were used in the interrupt should have been saved on the stack and restored before returning.

When an interrupt occurs, the processor saves the return address on the stack (usually). (The stack was described in Chapter 3.) The usual practice for entering an ISR is to save all the processor registers, just like any other subroutine. But, as Chapter 3 discussed, limitations on stack size sometimes make this impractical.

An ordinary (non-ISR) subroutine can be called with the knowledge that some registers will be changed during execution. An ISR cannot do this, as there is no way to know what registers are being used by the polling loop or a higher-priority ISR when the interrupt occurs. The ISR must save all registers that it uses, unless the software engineer can be *sure* that a particular register is used nowhere else. This is a dangerous practice in most cases, although there is sometimes no other choice

In addition to limiting stack size, some processors make it impossible to save registers on the stack. The PIC17C42, for example, has a 16-level stack and no PUSH or POP instructions. The stack is used for return addresses only. Values that must be saved, such as register contents, have to be stored in discrete RAM locations, as described for subroutines in Chapter 3.

In processors that have a hardwired stack, you must limit the levels of ISRs, or the stack will overflow. You cannot pass parameters on the stack (at least not very many) for the same reason.

The context switching registers described in Chapter 3 can be used for interrupts. For example, the 8051 has four independent register banks. One could be used for the polling loop, one for subroutines, and two for interrupts. For processors that have this capability, remember that common registers must be saved just like in a subroutine. Going back to the 8051, for example, there is only one accumulator register, so the ISRs must push it onto the stack. Of course, any register save method that works for a subroutine will work for an ISR.

Nested Interrupts

As mentioned earlier, some processors allow interrupts to be nested, which allows an ISR to itself be interrupted by another interrupt. The simplest method of interrupt nesting is to allow any ISR to be interrupted by any other. The other, more complex, method is to allow an ISR to be interrupted only by a higher-priority interrupt.

Interrupt nesting normally is used when a high-priority interrupt cannot wait. Without nesting, the lowest-priority interrupt becomes the highest while it is executing. If your design requires interrupt nesting, there are some special considerations.

The first consideration is the size of the stack. That becomes important if interrupts are nested. If you have an eight-level stack and nine levels of interrupts, you have an obvious problem.

Context switching, as mentioned earlier, can be a problem, depending on how many levels you have. I have designed several boards using the ADSP-2101 family parts, and all these designs use the alternate register set for interrupts. On these designs, I needed no interrupt nesting, so I always know that the alternate set is available to an ISR when it is executed. By definition, any previous ISRs have completed before the next one begins.

With or without nesting, interrupts pose a potential problem with timing. If all possible interrupts occur at once, the time before the polling loop can resume processing is the sum of the execution times of all ISRs. This seems obvious, but it is surprising how often systems are designed on the implicit assumption that all the interrupts will not occur at once. In a real-time system with asynchronous inputs, it is virtually guaranteed that this will happen eventually. The only exception is where you know some interrupts are mutually exclusive. For example, if you are using a half-duplex serial interface, you know you will not have receive and transmit interrupts simultaneously.

Passing Data to or from the ISR

When it occurs, an interrupt is an asychronous event. The polling loop can be doing anything, executing any instruction, at the time. Unlike a subroutine call, the ISR cannot pass information to the polling loop in a register, unless that register is unused by the polling loop *and* by all other interrupts.

The usual method of passing information from the ISR is via buffers and semaphores. The pool timer, for example, uses the following flags to pass information from the interrupt to the polling loop:

Flag	Function
ONFLAG	Set when the ON key is pressed by the user
OFFLAG	Set when the OFF key is pressed by the user
SEFLAG	Set when the SET key is pressed by the user
FCFLAG	Set when the FCN key is pressed by the user
MTFLAG	Set when the water low switch is closed
TFLAG	Set when time rolls over to 0:0

The interrupt routine sets these flags and the polling loop resets them after they are recognized. The pool timer does not pass specific parameters from the polling code to the ISR, but the ISR does use values set by the polling loop: the initial ON and OFF times, loaded when time rolls over, for example.

Some Real World Dos and Don'ts

Stuck Interrupts

As mentioned earlier, a stuck level-sensitive interrupt can cause the processor to hang up in the ISR. It is possible to detect a stuck interrupt (say, so you can shut off the motors when it occurs). If you know that the interrupt rate is slow enough that the polling loop should execute several times between interrupts, set a flag on each pass through the polling loop. The ISR checks the flag, and if it is not set, the interrupt must be stuck. If the interrupt can occur more than once for each pass through the polling loop, add a counter to the ISR and increment the count each time that the ISR executes when the flag is not set. If the count ever reaches a value higher than should occur in normal operation, the interrupt is stuck. With either approach, be sure to take into account other ISRs that may delay execution of the polling loop.

The Shared Memory or I/O Trap

Look at the following numbered lines of pseudocode:

1. Read location *xyz*.
2. OR the value 01 with the data from *xyz*.
3. Store the result back at *xyz*.

Now look at the following interrupt code:

```
OR the value 20h into location xyz.
Return.
```

If *xyz* contains, say, 00, and the first code (steps 1–3) is executed, followed by the interrupt, the result will be 21 in *xyz*. If, however, the interrupt occurs while the processor is between lines 1 and 2 or 2 and 3, the result will be 01 in *xyz*. The operation performed by the interrupt is overwritten by the code. You can see this if you rewrite the original psuedocode with the ISR code in the middle:

1. Read location *xyz*. (CPU reads 00)
2. OR the value 01 with the data from *xyz*. (Data = 01, but not stored in *xyz* yet.)

   —Interrupt occurs here—
   OR the value 20h into location *xyz*. (This is ISR code, ISR leaves 20h in *xyz*.)
3. Store the result back at *xyz*. (CPU now stores 01 in *xyz*. ISR operation is overwritten.)

This is a contrived example, but I have seen this exact scenario occur with an I/O register more than once. Avoid this pitfall. It can be extremely difficult to find, as it can be very intermittent. This leads to what I call the first rule of interrupts:

Wherever possible, avoid having interrupts that write memory or I/O locations that are also written by the polling loop or by other interrupts. Locations written by the polling loop should be read by the ISR and vice-versa.

The exception is certain semaphores. The pool timer ISR, for example, sets semaphores when a key is pressed. If the polling loop does not clear the semaphore (if it is in a mode where that key is ignored), the ISR resets the semaphore when the key is released. This is a "safe" violation of the rule, since the key press will never be so fast that the polling loop misses it. It is safe to violate this rule on occasion, but be sure you know it is really safe. Again, problems in this area can be very hard to find.

In cases where you must violate this rule because of hardware constraints (an I/O expander IC shared between the polling loop and the interrupt code, for example), disable interrupts before the write operation and reenable interrupts after the write. This will keep an ISR from altering the contents in the middle of a write. The following is the original pseudocode sequence, bracketed by the disable/enable:

Disable interrupts.

1. Read location *xyz*.
2. OR the value 01 with the data from *xyz*.
3. Store the result back at *xyz*.

Enable interrupts.

This, by the way, is the reason why common registers such as the 8051 accumulator must be saved in the ISR. If the polling loop just did an AND operation on the accumulator to check a bit and then the ISR changes the accumulator, the polling loop would make the wrong decision.

One way to avoid problems with shared hardware is to have a pair of mask bytes. For example, say an 8-bit register or output port is written by the software. The lower 4 bits are connected to status LEDs and the upper 4 bits turn four solenoids on and off. Let us also say that the polling loop controls the LEDs and an ISR controls the solenoids. This is an obvious case where a potential conflict can occur.

One solution to this is for the polling code to have a mask byte in RAM. The polling loop turns bits on and off in the LED mask but never writes to the hardware register. The ISR writes to the register, ORing the LED mask byte into its own solenoid control value. This could be reversed, of course, with the polling loop controlling the hardware and the ISR having a solenoid mask byte.

A related problem can occur on many peripheral ICs. The increasing complexity of modern peripheral parts requires them to have a number of internal registers. Rather than using a large block of processor address space, these parts sometimes have two addresses: register select and data. In operation, the processor writes the number of the internal register it wants to read or write to the peripheral's register select address. It then reads or writes the data address to modify the selected internal register. If a process in the polling loop or in a lower-priority ISR tries to access such a peripheral and an interrupt occurs between register selection and data transfer, and if the ISR also uses the peripheral IC, the original process will access the wrong register when it regains control. This happens frequently. It is an easy problem to overlook, and it is not always possible to avoid having the ISR and another task access the same peripheral IC.

Race Condition

A software race condition exists when a process tries to use data before the data really are ready. Say that a system has a process in the polling loop and a regular interrupt ISR. The polling loop process gets data from some external device such as an A/D converter or maybe commands via an RS-232 link to an external controller. The polling loop process places the input data in a memory location, DATABYTE. It also sets a flag (semaphore) byte, FLAGBYTE, to tell the ISR data are available. Each time the ISR executes, if the flag byte is set, it reads the data from DATABYTE and does something with that data, then resets the flag byte. Now look at the following code description for the polling loop process:

1. If input data are available,
2. Set FLAGBYTE,
3. Store input data at DATABYTE.

If the interrupt occurs between steps 2 and 3, the ISR sees the flag set and processes the data. But, since the polling loop process has not yet written the data, the ISR uses whatever value is in DATABYTE, which probably is the *previous* data value. In addition, the actual data byte is never processed since the flag was reset by the ISR.

The simple fix for this problem is to swap lines 2 and 3 in the code description or disable interrupts around those lines of code. Note that an optimizing compiler (see Chapter 3) can create a race condition under the right circumstances.

Cumulative Time Errors

Imagine a system where a timer generates an interrupt. Inside the ISR, the software reloads the timer for the next interrupt. This might be done because the timer, for whatever reason, is incapable of generating a regular output or because the time between interrupts needs to vary with external events. Look at the following ISR code description:

ISR entry.

Save registers on stack.

Calculate new timer value.

Store value to timer.

When the timer interrupt occurs, a varying amount of time will pass before the ISR actually is executed, depending on what the CPU is doing. If a higher-priority ISR is executing or if interrupts are not nested, then the variation in this delay can be quite large. Figure 4.7 illustrates this situation.

In the figure, the CPU loads and starts the timer at the beginning of each crosshatched area. The timer generates an interrupt and stops counting at the end of the cross-hatched area.

Assume that the delay between assertion of the interrupt and execution of the instruction in the ISR that loads the counter is 300 microseconds (µs) (area A on Figure 4.7) and the next interrupt has to occur 10 milliseconds (ms) after the current one. Since it takes 300 µs to load the counter, the time before the next interrupt actually will be 10.3 ms. This error accumulates; each interrupt interval is off by 300 µs. More important, since other factors can cause this value to vary, you cannot just subtract the error from the timer load value. For example, the delay indicated by area B on Figure 4.7 might indicate a case where the normal delay is increased because the CPU was doing something (such as a DMA transfer) that had precedence over the interrupt. The third delay, area C, might be where the CPU was executing another ISR when the timer interrupt occurred.

In this particular case, the best solution is to use a timer that can generate regular interrupts. A more likely scenario where that would not be possible is if the time between interrupts has to vary. In that case, this particular problem can be fixed by using a timer with a holding register that can be loaded with a new count while it is running and that starts using the new count

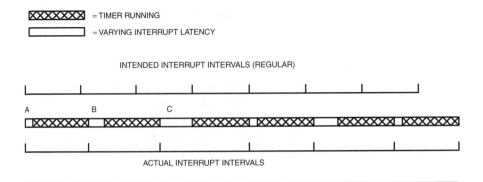

Figure 4.7
Variation in Interrupt Servicing (Latency).

when the current count expires. When the interrupt is generated, the timer automatically starts the next count and the ISR calculates the following count and loads it into the timer's holding register.

This example concentrated on timers, but the basic principle applies to any system where the software causes a timed value to vary. The basic rule is that, if a cumulative error is unacceptable, then the timing function (or at least the timebase) must be performed in hardware. An example of an area where cumulative timing error is no problem is an RS-232 interface implemented in software. Since the receiver resynchronizes on each byte, a small error in bit-to-bit sample timing is acceptable. On the other hand, in a time-keeping application, where time of day has to be maintained, an error of 1 count in 10,000 results in the time being off by a full minute at the end of a week. This is unacceptable in most such applications.

The time accumulation rule does not preclude software operation of a real-time clock. It just means that the timebase, which generates the regular time-keeping interrupts, must be maintained in hardware. This latency in servicing interrupts limits the precision with which software can perform any timing operation and must be taken into account in the design.

Multiple Reads

Suppose our polling loop has a piece of code that does this:

```
Read location X
If X = 0, do something
Read location X again
If X = 1, do something different
Read location X again
If X = 2, do a third thing
and so on . . .
```

The intent is that only one operation will be performed for each pass through the loop because X is not expected to change. Then suppose that an interrupt occurs between the first and second read and that the ISR changes the value of X. The result could be a path through the decision tree that appears impossible when you try to debug it. In cases where an ISR can change a memory location, never assume it will be the same on two separate reads outside the ISR. If the polling loop (or another ISR, if nested interrupts are used) needs to use the value twice, read it once, then store it in a register or a temporary memory location for subsequent use:

```
Read location X
Store it in location Y
```

If Y = 0, do something
Read location Y again
If Y = 1, do something different
Read location Y again
If Y = 2, do a third thing
and so on . . .

Now changes to X do not affect what the loop does. This scenario may seem unlikely, but remember that some simple microcontrollers have no compare-with-immediate-data instruction. On these processors, the only way to test for a specific value is to exclusive-OR the variable (X in this case) with the desired value and check for a 0 result. Then X has to be read again for the next test.

Minimizing Low-Priority Interrupt Service Time

Figure 4.8 shows a dual-DSP system, implemented using a pair of Analog Devices ADSP-2101s. The ADSP-2101 has a high-speed serial interface that can be used to communicate between two DSPs. In this system, which is a simplified diagram of an actual design, both DSPs had to service a regular, very high-priority interrupt every few microseconds. DSP1 controlled some high-speed (proprietary) circuitry. DSP2 processed the output from that circuitry and communicated with a host processor on another board. The two DSPs communicated via the serial link, sending 16 bits of data at a time.

I wanted to use the high-speed serial link for interprocessor communication, but processing the regular interrupt could not be delayed more than a

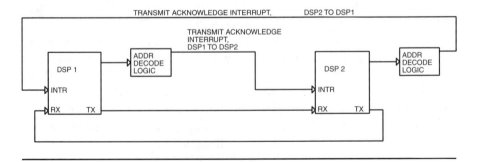

Figure 4.8
Dual DSP System with Communication Interrupts.

few clock cycles. Since the ADSP-2101 has only one alternate register set, which I already was using, I did not want to use nested interrupts.

The normal method of servicing the receive interrupt would be to read the word from the input register, check for errors, put the received word in a buffer, and then return. I could not afford to delay this long before servicing the regular interrupt.

One detail about the design that made the problem solvable was that the serial interface was not receiving or sending a continuous stream of data. Information was sent over the serial link intermittently.

The solution was to add an edge-sensitive acknowledge interrupt from DSP1 to DSP2 and vice-versa. When a byte is received by DSP2 from DSP1, the receive interrupt code just sets a flag and returns. The polling loop checks the flag once per pass, and if data are available, it reads and processes the received word.

The acknowledge interrupt is used to prevent data overruns if DSP1 wants to send more than one word. When DSP1 transmits, it clears a semaphore, TXRDY. It will not transmit again until TXRDY is set, which happens when DSP2 sends back the interrupt. DSP2 does this when it has processed a received byte. If the polling loop gets busy processing host data and does not get to the received byte, DSP1 just waits. If necessary, the FIFO buffer can continue to fill with pending transmit data, although this rarely occurred in the actual system.

Since the acknowledge interrupt and receive interrupt merely set flags, they need not save any registers. The ISRs consist of two instructions: one to set the flag and an interrupt return.

The pseudocode for the polling loop and interrupts follows (DSP2 is shown; DSP1 is the same without the host processing part):

```
Start of polling loop:
If RXRDY is set (Rx serial data available),
        Read data byte
        Send acknowledge interrupt to DSP1
        Process Rx data.
If host data available,
        Read host data
        Process host data.
If transmit data available for DSP1,
        Write data to transmit FIFO buffer.
If transmit FIFO has data,
        If TXRDY flag set,
                Reset TXRDY flag
                Transmit a word to other DSP.
```

```
End of polling loop.
        Rx serial data ISR:
                Set RXRDY
                Return.
Transmit acknowledge ISR
                Set TXRDY
                Return
```

When to Use Interrupts

This topic almost makes more sense at the beginning of the chapter, but I wanted to talk about how interrupts work before discussing why they are not always a good idea.

To begin with, sometimes there is no choice but to use interrupts. The previous dual DSP, for example, had to use an interrupt for receiving, because there is no internal status bit that can be checked by the polling loop to see if data are available.

Chapter 1 stated that designers sometimes overuse interrupts. Although it is dangerous to lay down universal rules like this, in general, interrupts should be used for three reasons:

1. To generate a regular, repeatable event. A timer interrupt on a regular basis for timekeeping is a good example.
2. When a peripheral absolutely has to be serviced immediately or something will go wrong; for example, a UART receiving a continuous stream of data that will overrun the receiver if each byte is not taken out before the next one is received. Be careful, though, about assuming that cases like this require an interrupt. The protocol converter described in Chapter 3 used no interrupts. The program loop was short enough and fast enough, even with subroutine calls, to guarantee that no received bytes were missed.
3. To save on hardware. For example, a processor might get an interrupt from the shaft encoder of a motor. Each time the interrupt occurs, the processor increments a position count. This saves adding a hardware counter. The pool timer uses an interrupt to multiplex the display, saving the cost of external display registers and multiplexing logic.

Interrupts also are used as a means of scheduling tasks through a real-time operating system, which will be covered in Chapter 8.

Remember, though, that interrupts make debugging more difficult. When you set a breakpoint using a monitor program or emulator, interrupt inputs keep coming. You cannot always single step code when using interrupts, because continuous interrupts keep the processor hung up in an ISR. And when something goes wrong because a high-priority interrupt did not get the priority it needed, the source of the problem can be very hard to find.

The polling loop is sometimes called the *background loop* because it is the lowest-priority task—everything else can interrupt it. The background is executed when no interrupts are being serviced. Designers sometimes assume that the polling loop therefore is the worst place to put anything that needs priority attention. But keep in mind that a low-priority interrupt may be just as bad as the polling loop. If servicing a device such as a UART in the polling loop means that the worst-case stack up of interrupt service times will prevent the device from being serviced in time, making the device a low-priority interrupt may be no better. Look carefully in a case like this to see if there is a basic throughput problem when interrupts stack up.

In some cases, a device such as a UART may have a momentary problem with interrupt service time stack up, but you know that this is a temporary and infrequent condition. In those cases, you may be able to solve the problem by using a buffer such as a FIFO buffer. Our example UART, with either an internal or external FIFO buffer, may receive several bytes before the polling loop gets to it. If you really (really, really) know that the problem is a temporary one, the polling loop should be able to process the data before the FIFO buffer fills up. But remember that, when the software gets to the UART, it will take longer to process the data since there are more bytes. Make sure the throughput problem really is temporary and that the polling loop will get through with its delayed processing before the next crunch occurs.

The software, hardware, and interrupt structure are the primary components of the embedded system. Chapter 5 explains adding hardware and software to simplify the debug process.

Adding Debug Hardware and Software

5

In designing the hardware and software for an embedded microprocessor system, we typically give most of our attention to the actual application. But all embedded systems have to be debugged sooner or later. The addition of selected hardware and software can simplify the debug process.

Almost any embedded system can benefit from the strategic addition of hardware and software for debugging. Depending on cost and size constraints, the hardware may not be installed but instead connected to the circuit only when data collection actually is required. But the software must support the functionality.

This chapter focuses on adding debug capability that allows operation of the software to be traced, to provide a history of what has happened when something goes wrong. Developers who cannot use an emulator, for whatever reason, will see the obvious utility of the techniques to be presented in this chapter. Users who have access to sophisticated emulators may question why additional debug tools should be incorporated into the design. There are three reasons: First, the equipment may have a subtle bug that shows up only in a customer's environment where an emulator cannot be connected or where use of the emulator would stop critical work. Some types of equipment cannot be stopped while the software engineer looks through the emulator traces to find what caused a problem. In cases like this, when an error occurs, the data are captured immediately and the machine must be restored to operating condition immediately. The data must be examined later.

A second reason for adding debug tools to the system is to give a macro history of system operation. Embedded real-time systems often control mechanical devices, which can have a significant lag between an error and the event that caused the error. An emulator can give very detailed information around the error itself, but its trace buffer often is not deep enough to

---

Some of the information in this chapter originally appeared in *Circuit Cellar Ink* magazine.

allow operation around the *cause* of the error to be examined. One real world example of this is a document processor that I worked on. This machine moves checks and other paper at a rate of over 1600 documents per minute. An 8031-based subsystem (described earlier in Chapter 3) was having problems controlling document spacing. The error could not be detected until about 100 milliseconds (ms) after the event that caused the error. On an 8031, 100ms is 100,000 instructions, and the emulator just could not store that much data. Fortunately, I did not need to see every instruction to determine what was wrong. Using a technique similar to the ones described in this chapter, I was able to capture just the pertinent information on a logic analyzer, and locate the actual problem.

A third reason for adding debug tools involves limited resources, such as those found in single-chip microcontroller systems. If you have only 2K of program space to work with, it may be difficult to fit a full debugger into the chip along with the application code. Simple debug tools let you see what is going on inside the code when you lack the room for a full development environment.

The key to adding debug tools to an embedded system is determining what to save. In general, I like to keep track of the following as a minimum:

Each entry to an interrupt routine.

Each exit from an interrupt routine.

Each time that a major function is performed (such as execution of a command from a higher level controller).

Each time that a message or command is passed to the outside world.

Finally, debug tools are useful as a means of correlating software execution with the real world as captured on a logic analyzer or digital storage oscilloscope (DSO).

There are many ways to trace this information. They center around two basic techniques: generating the data for hardware that may or may not be connected and storing the information in memory for later retrieval. Both techniques have advantages and drawbacks.

Hardware Output

The hardware output technique depends on getting debug data to the outside world in some simple 8-bit, 16-bit, or even 32-bit format. Ideally, the firmware will not check to see if the debug hardware is connected, because reliable debug depends on the firmware operating the same whether the debug hard-

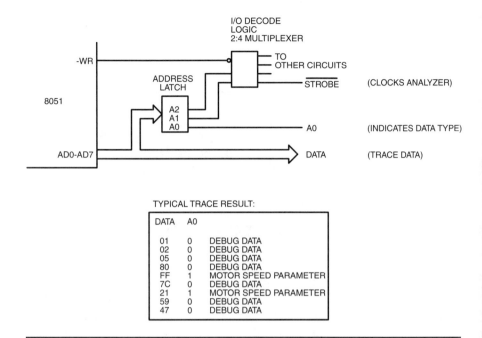

TYPICAL TRACE RESULT:

DATA	A0	
01	0	DEBUG DATA
02	0	DEBUG DATA
05	0	DEBUG DATA
80	0	DEBUG DATA
FF	1	MOTOR SPEED PARAMETER
7C	0	DEBUG DATA
21	1	MOTOR SPEED PARAMETER
59	0	DEBUG DATA
47	0	DEBUG DATA

Figure 5.1
Example of 8051 Trace Output Logic.

ware is connected or not. In the 8031 example mentioned earlier, the processor used external read-only memory (ROM) and I/O ports. I connected an unused I/O decode to the clock of the analyzer and captured the 8-bit 8031 data bus as illustrated in Figure 5.1. The firmware was modified to write data to the I/O port when certain events occurred. Using the logic analyzer, the sequence of events and the time between them could be examined. A typical table of output values for a document processing subsystem like the 8031 system might look like Table 5.1.

This is just a sample of the sort of thing I collected for that problem. The actual table of output data was longer, had more interrupt points, and some of the actual debug definitions would make little sense here without a lengthy explanation of how the code worked. The significant events are indicated by an output to the data port. Note that not every instruction or even every subroutine call is captured, just major events.

Interrupt entry and exit are captured so the time spent in the interrupt routine can be measured. In Table 5.1, the interrupt routines use data values of 80h and above. This is because most logic analyzers can be programmed to ignore data using a binary mask. To capture everything, no mask is used. If I know the interrupts are not involved in the problem, I can

Table 5.1
Debug Output Values for Typical Embedded System.

Value (hex)	Meaning
01	Throughput too high—ramping down
02	Throughput too low—ramping up
03	Document detected at sensor 1
04	Document detected at sensor 2
40	Document spacing error detected
80	Timer interrupt entry
81	Timer interrupt exit
82	Command interrupt entry
83	Command interrupt exit

program the analyzer to ignore any data bytes where bit D7 is set. The interrupt entry and exit codes are not stored so there is more room in the buffer to capture all the other codes. In the actual situation, I did exactly this—I captured everything until I could verify that the interrupts were not a factor, then I ignored the interrupt trace values after that.

One of the tasks this 8031 performed was to control the speed of a motor. I wanted to monitor certain parameters about the motor speed control algorithms as well as capturing the debug data. Unfortunately, the data could be any value, so I could not mix it with the trace debug data. I had have no way to tell the difference between data and trace codes.

Since the I/O decoding logic used partial address decode (see Chapter 2), I actually sent debug data to two different port addresses, specifically 3 and 4. Address 3 got the data as just described and the address 4 got the motor speed parameter. Both port addresses activated the I/O strobe that generated the capture clock, but A0 was different for the two addresses. By connecting A0 as an input to the logic analyzer, I was able to tell the data type of each captured byte.

This lengthy discussion about a specific 8031 system illustrates the kind of trace information that can be captured using this type of debug method. Other methods can be used to generate the trace information as well.

Write to ROM

In some systems, there are no spare I/O decodes. If the processor uses external ROM, it often is possible to write debug information to the ROM space.

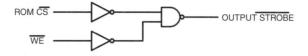

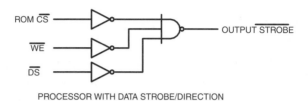

PROCESSOR WITH DATA STROBE/DIRECTION

Figure 5.2
Logic for Write to ROM.

Most systems do not enable the ROM when it is written to, so there will be no bus conflict, and usually nothing else is in the ROM address space. The circuit in Figure 5.2 shows write to ROM circuits for both Intel-type (RD/WR) buses and Motorola-type (strobe/direction) buses. A low-going strobe is generated when the firmware writes anything to the ROM space. This technique is useful when all the available I/O decodes are used up and trace capability needs to be added after the hardware is finalized. When needed, the decoding logic can be connected to the board via a DIP (dual inline package) clip or PLCC (plastic leaded chip carrier) extender. The one drawback to this method occurs in systems that have flash or other writable memory, where a careless write sequence actually could change the data.

Like the write to an I/O port, this method can use different addresses for different data types. Note that, if the Motorola-type bus is used, a DTACK must be generated for the debug write cycle.

Read from ROM

Some microcontrollers, such as the 8051, lack the capability of writing to the ROM space. However, they do have instructions that can read from ROM to do things like table lookups. Figure 5.3 shows a circuit that can be used to

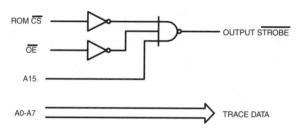

Figure 5.3
Generating Trace Data by Read from ROM.

generate trace data by this method. A low-going strobe is generated when a read is performed from any PROM address above 8000h. When using this circuit, the trace data is captured from the low-order *address* lines, A0–A7, and the data lines are ignored. The circuit shown in Figure 5.3 assumes that the ROM is located in the lower 32 K. If the ROM is larger than 32 K, a wider NAND gate can be used to further decode the address bits, generating a strobe only for the upper 1 K or less. The following code shows how this would be implemented on an 8051 family microcontroller:

```
MOV DPTR,#8000h ; starting address—upper 32k
MOV A,#CONSTANT ; constant is the trace value
MOV A,@A+DPTR ; output the value.
```

This method can use different addresses to identify different types of trace data, but both addresses must be in the upper 8 bits of the address field, since the lower bits are used for the data. For example, you might use address 8000 for trace opcodes and C000 for data.

Software Timing

Although many emulators can perform this function, the tracing technique described earlier also can be used for software timing. The software returns a trace value at the entry and exit of each major subroutine, function call, or whatever level of code is to be analyzed. Using a logic analyzer with timetagging capability, the trace values are captured; and you can determine how long the code spends in each routine. I used this technique when working with a software engineer on a problem where the software would just "go away" for a while. It did not lock up; it just took too long and other things quit working because of the delay. We added trace points to the software and discovered that the problem occurred in a function that parsed data received from another system. When the error occurred, we clearly saw a trace point generated from entering the routine and the exit from that routine occurring much too late.

Of course, this technique also can be used to determine how much time the software is spending in each routine or how long it spends servicing each interrupt.

Software Throughput

Sometimes it is necessary just to have an idea of how busy the processor is while it is running, to know if there is a problem with throughput. One technique is to set a port bit each time an interrupt or time-critical task occurs, then have it continuously reset by the idle loop. If interrupts are nested (see Chapter 4) or if one time-critical task calls another, the bit will be set twice, which would not affect it. The idea is to get the bit to be low only when the code is in the idle loop. The ratio of high to low time, as seen on an oscilloscope, gives a fairly good representation of the utilization. The changing ratio can be monitored while the system runs.

Circular Trace Buffers

Sometimes it is impractical to add debug hardware or trace in real time. In these cases, some trace capability can be implemented with a rotating trace buffer. This is an area of memory written by the software with trace data, similar to the hardware trace function.

The data area is configured as a circular buffer. When data are written to the last location, the pointer wraps around to the first position. The size of the buffer depends on how many trace points are needed and how far back the history needs to go. Table 5.2 shows a short (16-byte) trace buffer that might be used to debug a simple system. The table shows 8-bit hex codes and comments indicating what each code is for.

In this example, the code writes a byte to the table for each trace value, then writes 0FFh to the next table location. When the last table location (000F) is reached, writing wraps back around to location 0000. In this example, location 0004 contains FF, so the last entry in the table is 01, timer interrupt entry, at location 0003. The history goes back from there. Note that

Table 5.2
Short Trace Buffer.

Address (offset into start of table)	Trace Value	Meaning
0005	01	Timer interrupt entry
0006	02	Timer interrupt exit
0007	04	Start of output processing
0008	05	Output processing done
0009	09	Command received from host
000A	07	Operator pressed key
000B	0A	Time rollover
000C	04	Start of output processing
000D	01	Timer interrupt entry
000E	02	Timer interrupt exit
000F	05	Output processing done
0000	08	RS-232 byte received
0001	01	Timer interrupt entry
0002	02	Timer interrupt exit
0003	01	Timer interrupt entry
0004	FF	Table terminator

the timer interrupt occurred between the start and end of output processing at locations 000C–000F.

There usually will be a table pointer in another location and it also can be used to keep track of the end of the table. Either way, the engineer or some software has to analyze the pointer or table values to find the end of the table.

If this technique is used, it is best implemented with a single subroutine that writes to the buffer and updates the pointer or pointers. When the pointer reaches the end of the buffer, it must be wrapped back to the beginning. Using a circular buffer means you do not have the data captured in a logic analyzer, where it could be correlated with real world events.

Data can be time tagged in the trace buffer if a clock is available. One way to do this is to have a free-running hardware timer and store the count in the trace buffer with each trace point saved. Some subtraction will determine relative times. Of course, the timer eventually will roll over, so you have to take that into account as well.

This technique is useful if the only access to the outside world is via a serial port, which is not fast enough to capture the trace points in real time. One caution though: If a regular time interrupt is used in the system, beware of adding it to the trace buffer. If an error occurs and the timer keeps running, it can quickly fill the buffer and the information you need will be lost.

Embedded Microprocessor Systems

A last note about trace buffers: Some software debuggers provide an event capture function that can be called from the application code. This function typically is passed information that is timetagged and stored in a trace buffer. These debuggers in effect, provide the circular buffer with very little extra work. Of course, the debugger software must be resident for this to work. In the final product, that can mean additional license fees.

Monitor Programs

A software monitor/debugger, mentioned in Chapter 1, is a program that usually resides in PROM and allows the user to examine memory, registers, and I/O ports. Many software engineers write their own monitor programs, although this is less common than it used to be. Most monitor programs have some basic capabilities in common:

Ability to examine and alter memory.

Ability to read and write I/O ports, if any.

Ability to set breakpoints in the code.

Ability to download code from a host PC.

Ability to interrupt code execution from the keyboard.

Communication via an RS-232 serial port, usually.

The monitor program generally gets control on power-up. In the early days, the user interface was a dumb terminal, but a dedicated PC is used universally today. The user generally will download code into the target system RAM and execute it there. The reason the code has to run out of RAM is so that breakpoints can be set. A breakpoint is a branch instruction that replaces the instruction that normally would appear at a particular location. The branch returns control to the monitor program.

To use a monitor program, the user puts a breakpoint at some useful place in the code, runs the program, and then examines memory and registers when the breakpoint occurs. Breakpoints can be placed in error routines to see if the error occurs.

Logic Analyzer Breakpoints

Since embedded systems are connected to the real world, you often need to know what the processor is doing when specific real world events occur. This may be because the event represents an error condition or just because you need to know how the software handles it (not how it is *supposed* to handle it, but what it really does). You may be able to capture the event on a logic analyzer or DSO, but how do you take a "snapshot" of what the software is doing at that time? In many cases, you can take advantage of the trigger output port of your logic analyzer or DSO. This typically is a BNC connector with a logic level signal that goes high or low, or pulses, when the analyzer or DSO triggers it.

If you are using an emulator, often there is an external input that you can breakpoint on. By connecting the trigger output of the logic analyzer or DSO to this external input, you can force a breakpoint.

If you are not using an emulator or have no external breakpoint input, you can stop the software by using an otherwise unused interrupt, using a timer as an interrupt (see Chapter 4), or even using an NMI. Another possibility is to use the reset input, but you need a means to keep the software from reinitializing everything after the reset. Using the reset technique will not let you see what the software does after the event, but it will let you look at the state of the software just before the event.

Finally, you can use this technique to capture the contents of a rotating trace buffer when a real world event occurs. If you use any of these techniques to connect a logic analyzer or DSO trigger to your system, be sure the trigger output is compatible. You may need to invert the polarity or change the pulse width to make everything work.

Memory Dumps

I once worked with a consultant who was interfacing an add-on to our equipment for a third-party reseller. My job was to make sure that nothing we did or failed to do would keep him from doing his job. This individual had an interface subsystem that consisted of dual microcontrollers communicating via hardware FIFO buffers. Every now and then, the entire thing would lock up for no apparent reason. He had no tools except a fairly simple logic analyzer and the stare-at-the-code method of debug. At that time, no good emulators were available for the processor he was using. The entire project got so far behind schedule that the reseller sent out its ace troubleshooter to get things going. The troubleshooter was aggressively pro-active and carried a custom-made wooden briefcase. He even went so far as to order, on the company credit card, via overnight delivery, an emulator that was supposed to be the best there was. That was when we found out nothing was suitable.

Embedded Microprocessor Systems

The consultant told me that if he could view the internal memory contents, he could fix the problem. We went over an inventory of what he had available, which was little. The board had no spare output ports but one unused interrupt input. We solved the problem by connecting a push-button switch to the interrupt. When the button was pressed, generating an interrupt, the processor produced as output three sync bytes in a sequence that would never occur in normal operation. Then the processor sequentially wrote the entire contents of the internal RAM to an external register. The logic analyzer, connected to the register, waited for these three bytes (FF FF FF, if memory serves) then triggered and captured everything that came afterward. The register contents were trashed by this procedure, but we had to reset everything after the lockup anyway so it did not matter. The analyzer, like most analyzers, numbered the states. Due to the triggering setup, the state number was the same as the memory address, so finding a particular byte was easy.

Another way to handle this, if a spare output port bit is available, is to write a serial output routine that would transmit the data over the port bit asynchronously (implementing a bit-banging transmitter in software). The data could be converted to RS-232 and captured on a dumb terminal or a PC.

Although examining memory may seem like a throwback to the days when mainframe computers produced what was known as a *core dump*, the technique can be very useful. In a simple microcontroller design, there usually is not all that much memory to examine and only certain locations are important. You can combine the memory dump with the logic analyzer breakpoint described earlier and capture the contents of memory when an external event occurs.

Serial Condition Monitor

Sending data to an external logic analyzer, typically, requires 8 data bits and a write strobe. In a microcontroller design, especially a small one, not enough I/O pins may be left over to collect diagnostic data. However, if one port pin is available, it can be used to generate diagnostic information serially. The processor sets the pin to 1, then serially sends a 4-bit diagnostic word. The diagnostic word typically is a byte in memory, with bits set or reset by various routines to indicate the state of the code. For example, D0 might be set to indicate the state of flow control (ON or OFF), D1 might indicate when a servo motor is at the home position, D2 tells when a particular buffer has data to send or receive, and so on.

To use this method, the processor will return the diagnostic once each time it goes through the polling loop or once each time a regular timer interrupt is serviced. The output data are captured on an oscilloscope or logic analyzer. An oscilloscope should be set to trigger on the first sync or start bit. (The start bit is needed so you can tell which of the other bits are set.) The variable sweep control can be adjusted so that each output bit takes one or two divisions. This method can be extended to any number of bits, but trying to do more than about four makes it difficult to determine which bits are set.

Figure 5.4 shows the waveform that would be generated using this method for data patterns of 10 (0A hex) and 11 (0B hex). The bits can be captured in a shift register by using logic that starts on the sync bit and samples in the middle of each bit.

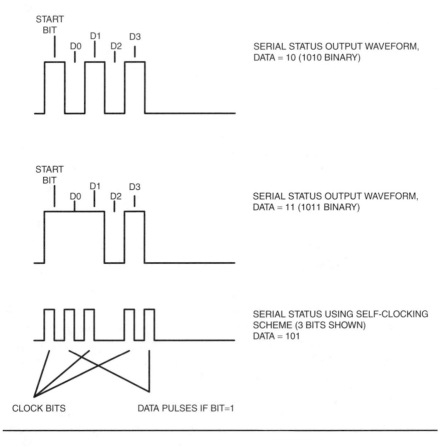

SERIAL STATUS OUTPUT WAVEFORM,
DATA = 10 (1010 BINARY)

SERIAL STATUS OUTPUT WAVEFORM,
DATA = 11 (1011 BINARY)

SERIAL STATUS USING SELF-CLOCKING
SCHEME (3 BITS SHOWN)
DATA = 101

Figure 5.4
Serial Condition Monitor for Use with Oscilloscope.

There are some limitations to using this method. Since the data is sent on a regular basis, it cannot be used to determine the relative time between processes or events. However, it is useful for getting a running pictorial view of overall system status.

I have used this technique to send up to 17 bits of status information, although it is not useful to display it on a 'scope. Instead, I built an external circuit, using a CPLD, which picks up the serial data and captures it in a shift register. When all 17 (or whatever) bits are finished, the circuit generates a strobe to clock the diagnostic word into a logic analyzer.

This technique also works with self-clocking data. In this scheme, each data bit starts with a clock bit, and then there is a pulse in the middle of the bit time if the bit is a 1, and no pulse if the bit is 0. Figure 5.4 also shows the waveform for this technique. This is useful if you cannot turn the interrupts off long enough to send an entire diagnostic word. Since each bit is self-clocking, you can turn interrupts on momentarily between bits without upsetting the external hardware.

The asynchronous data can be received with a UART, either a discrete UART IC (if you need only 8 bits of data) or a UART implemented in a CPLD. I usually use a CPLD, since I normally send more than 8 bits when I use a circuit like this. If you use a UART, you need not run at a standard baud rate—you can give it a clock that matches whatever speed the processor is capable of sending. But keep in mind that some UARTs cannot be clocked fast enough to keep up with the data if you use a fast processor.

Figure 5.5 shows a block diagram for a circuit that will receive the self-clocking serial diagnostic output scheme. I implemented a circuit like this in a CPLD. The frequency of the reference clock depends on the type and speed of processor you use. The diagram does not show the power-up reset logic. The size of the bit counter and the output shift register depend on the number of bits transmitted from the processor you are testing.

Listing 5.1 shows assembly code for a PIC17C4x that generates the serial diagnostic data to Port D, bit 0. Listing 5.2 shows similar code for an 8051 using Port 0, bit 0. The 8051 version is simpler to implement, since it has an instruction that allows the ALU carry status bit to be written directly to a port pin. Listing 5.3 shows the code to produce a 16-bit, self-clocking diagnostic string using a PIC processor, sending to Port B, bit 7.

Of course, any of these code snippets could be converted to C or another language, but in some cases the timing will change. That depends on the compiler you use.

Listing 5.1. PIC17Cxx Assembly Code for Serial Status Output

```
; STATUS OUTPUT TO PORT D BIT 0.
; PORT D BIT 0 MUST BE CONFIGURED AS OUTPUT.
```

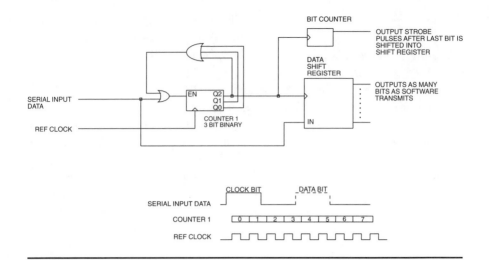

Figure 5.5
Circuit to Receive Data from a Self-Clocking Serial Diagnostic Output.

```
; CODE SENDS LEAST SIGNIFICANT FOUR BITS OF
; BYTE DIAG TO PORT.
; OUTPUT SEQUENCE IS:
; SYNC BIT (1)
; DIAG BIT 0
; DIAG BIT 1
; DIAG BIT 2
; DIAG BIT 3
; ZERO
; EACH BIT TAKES ABOUT 1.6 MICROSECONDS ON A PIC17C43
; WITH A 12 MHZ CRYSTAL.

      MOVLB 1
      BSF PORTD,0 ; OUTPUT SYNC BIT (1)
      NOP
      NOP
      BTFSC DIAG,0 ; DIAG BIT 0 SET?
      GOTO MSET0 ; YES, OUTPUT A 1.
      BCF PORTD,0 ; NO, OUTPUT A 0.
      GOTO M1
MSET0: BSF PORTD,0
      NOP
M1: BTFSC DIAG,1 ; DIAG BIT 1 SET?
```

Embedded Microprocessor Systems

```
        GOTO MSET1 ; YES, OUTPUT A 1.
        BCF PORTD,0 ; NO, OUTPUT A 0.
        GOTO M2
MSET1: BSF PORTD,0
        NOP
M2: BTFSC DIAG,2 ; DIAG BIT 2 SET?
        GOTO MSET2 ; YES, OUTPUT A 1.
        BCF PORTD,0 ; NO, OUTPUT A 0.
        GOTO M3
MSET2: BSF PORTD,0
        NOP
M3: BTFSC DIAG,3 ; DIAG BIT 3 SET?
        GOTO MSET3 ; YES, OUTPUT A 1.
        BCF PORTD,0 ; NO, OUTPUT A 0.
        GOTO MSDON
MSET3: BSF PORTD,0
        MSDON: NOP
        NOP
        BCF PORTD,0 ; TERMINATE WITH A 0.
        MOVLB 0 ; END OF DIAGNOSTIC OUTPUT.
```

Listing 5.2. 8051/8052 Assembly Code for Serial Status Output

```
; THIS CODE FRAGMENT OUTPUTS A FOUR-BIT STATUS
; VALUE, DIAGNOSTIC, TO PORT 0 BIT 0 OF AN
; 8051/8052 PROCESSOR. OUTPUT SEQUENCE IS:
; START BIT (1)
; DIAGNOSTIC BIT 0
; DIAGNOSTIC BIT 1
; DIAGNOSTIC BIT 2
; DIAGNOSTIC BIT 3
; ZERO.
; ON AN 8051 WITH AN 8 MHZ CRYSTAL,
; EACH BIT WILL BE ABOUT 4.4 MICROSECONDS LONG.

        MOV ACC,DIAGNOSTIC
        SETB P0.0 ;      OUTPUT START BIT (1)
        RRC A
        MOV P0.0,C ;     OUTPUT DIAG BIT 0
        RRC A
        MOV P0.0,C ;     OUTPUT DIAG BIT 1
        RRC A
        MOV P0.0,C ;     OUTPUT DIAG BIT 2
```

```
          RRC A
          MOV P0.0,C ;        OUTPUT DIAG BIT 3
          NOP
          NOP
          CLR P0.0 ;          ALL DONE
```

Listing 5.3. PIC Assembly Code for Self-Clocking Serial Status Output

```
; ------------------------------------------------------------------
; Synchronous serial diagnostic output on PIC.
; Sends 16-bit diagnostic word in DIAGLSB/DIAGMSB,
; to RB7. LSB of DIAGLSB is sent first, MSB of DIAGMSB
; is sent last.
; Output is clock bit followed by data clock if bit
; was a '1', no data clock if bit was a '0':
;                _   _
; 1: __/ \__/ \__
;                _
; 0: __/ _____
;
; This makes the data self-clocking.
; Interrupts are momentarily enabled between bits
; to minimize latency time.
; ------------------------------------------------------------------

diagnostic:

          movlw 8             ; bit count, LS byte
          movwf txbits        ; store at bit counter

          ; Transmit LS byte first
diaglsblp:
          bcf intcon, gie     ; disable interrupt during bit
                              ; output time.

          bsf portb, 7
          rrf diaglsb,f       ; rotate DIAGLSB.0 into cy
          bcf portb, 7        ; clock bit falling edge
          btfsc status,c      ; check carry
          bsf portb,7         ; set data clock if bit = 1
          nop
          bcf portb, 7        ; clear data clock, set or not.
```

```
        bsf intcon, gie          ; enable interrupts again

        decfsz txbits,f          ; done?
        goto diaglsblp

        movlw 8                  ; bit count, MS byte
        movwf txbits             ; store at bit counter

        ;Transmit MS byte

diagmsblp:

        bcf intcon, gie          ; clear interrupt during bit
                                 ; output time.
        bsf portb, 7             ; clock bit rising edge
        rrf diagmsb,f            ; rotate DIAGMSB.0 into cy
        bcf portb, 7             ; clock bit falling edge
        btfsc status,c           ; check carry
        bsf portb,7              ; set data clock if bit = 1
        nop
        bcf portb, 7             ; clear data clock, set or not.

        bsf intcon, gie          ; enable interrupts again

        decfsz txbits,f          ; done?
        goto diagmsblp

        return


; -------------------------------------------------------------------------------
; End of diagnostic output code
; -------------------------------------------------------------------------------
```

Now that we have all the tools, in Chapter 6, we look at the process of integrating the software and hardware to produce a working system.

System Integration and Debug   *6*

After the hardware PC board or a handwired prototype is built and the software compiles with no errors, it is time to plug in everything and watch the system come to life. Fifteen minutes after initial power-up and flawless operation, the hardware and software engineers knock off early for the day.

Well, not really. The actual scenario usually goes more like this: If smoke does not roll out when power is first applied, the engineers quickly discover that absolutely nothing works. After a brief argument where the hardware engineer accuses the software engineer of forgetting to initialize the stack, a logic analyzer or emulator is connected. Investigation reveals that the PROM or the RAM or something else appears to be completely dead. The software engineer goes to lunch, smug in the assurance that it is a hardware problem. After the hardware engineer fixes the chip select that had two devices enabled at the same time or the wiring error that had the data bus reversed end for end, debugging resumes. When the next problem stops all work, it turns out that an interrupt vector is in the wrong place or the internal processor peripheral area is uninitialized, and the hardware guy leaves early while the software engineer works until 8 P.M. to fix it.

This process repeats for the next four weeks as the project stumbles its way to completion. Two weeks after the design team breathes a sigh of relief and hands it off to the beta test customer, a bug is reported. The error occurs only under certain conditions and not every time at that. If the design team is lucky, it can reproduce it in the lab and work on it there. Otherwise, another argument ensues as to who has to take a midnight flight out to the beta site to fix it—hardware, software, or both?

While the utopian fantasy I first described is rare, the nightmare scenario need not be the only alternative.

Like the design part of a project, the test and integration part works best if there is a goal—a plan. Many large companies have formal test plan requirements that detail every test to be performed. These range from functional tests to safety agency approval tests to tests involving shock, vibration, and temperature (called by some *shake and bake*).

The market- or customer-specific tests are outside the scope of this book. This chapter concentrates on the functional test side.

Before product functional testing can begin, the hardware and software must be made to work. Integration of the software and hardware in an embedded system is different from a completely hardware (logic) or completely software (as an application on a PC) design in two areas: isolating problems and fixing them.

In isolating problems, it is not always clear in which half of the system the problem lies—hardware or software. In a pure software design, the programmer knows that hardware problems do not affect the design—they just mean that the computer needs to be fixed before work can proceed.

This may seem obvious, but I have seen software engineers used to writing applications for a minicomputer or PC who just walked away when a problem was thought to be in the hardware. They were used to working on software-only systems. Hardware bugs were a job for the repair technician. In an embedded system, the software and hardware engineers usually have to work together far enough into the problem to determine what actually is happening. Many times a hardware problem that seems completely mysterious will be obvious when the software engineer tells the hardware engineer just exactly which status bit the software is having problems with.

Fixing problems in an embedded system has differences, too. For example, sometimes the fix for a hardware problem is best implemented in software. An example might be a mistake in the schematic that places a peripheral at the wrong address. The board could be altered, but that requires another layout/fab cycle. It is simpler just to change the software. The reverse also is true, although less common.

Hardware Testing

Ideally, the software will not be tried until the hardware is completely checked out. In the real world though, that often is impractical. A decision must be made as to how the hardware will be tested, and how far the testing will go before software is installed. In a real system, where the software controls real hardware, the hardware cannot be checked without some software to exercise it. Look at the hardware checkout list for the pool timer:

- **Test processor and EPROM.** Write code that just loops, toggling a port bit on each cycle.

- **Test switches.** Write special software to read switch/water low connector and echo result to Port 1. Test each switch and corresponding port bit by verifying that correct Port 1 bit goes active when switch is pressed.
- **Test display.** Write special software that displays an incrementing number on both pairs of display digits. This requires working decimal-to-seven-segment code.

After the software is completed, the interrupt processing time and overall throughput are verified. This is done by setting a spare port bit at the start of interrupt processing and resetting the bit when interrupt processing was completed. An oscilloscope is used to verify the timing. Note that each of the hardware tests requires special software. In this case, the software is simple, but that will not always be true. At some point, writing special test software becomes less effort than just plugging in the real software and trying it out.

If an emulator is used for software debug, it can simplify hardware debug as well. Checkout of the pool timer hardware can be accomplished by plugging in the emulator, then verifying that the correct port bit changes when a key is pressed. The motor relay can be checked out by manually setting and resetting the corresponding port bit. A UART could be checked by wrapping input to output, programming the baud rate and other parameters, and checking that characters written to the transmit register are echoed in the receive register.

Deciding what to test and what software is needed can be part of a comprehensive test plan or it can be a list made up by the hardware engineer over lunch on a simple design. Most designs have no comprehensive list of what specific hardware will be tested unless software engineering support is required for the special code.

Software Debug

Keeping with the divide-and-conquer strategy of debugging, we look at the process I used to debug the software in the protocol converter described in Chapter 3.

Connect the serial side to a PC. Send a continuous string of characters when activated by grounding a spare port bit. This verifies that the setup code for baud rate, parity, and other settings works correctly.

Add code that handles serial I/O. Add special code that echoes serial input to output. Verify, using the PC communication program, that echo works correctly for all baud rates and parity selections.

Add proprietary processing code and verify that the echoed data are correct.

Load special code to send a test string to an output device using the final output driver.

Finally, integrate the setup code, serial I/O code, processing code, and output code. Verify correct functionality and XON/XOFF operation.

This a somewhat extreme breakdown for a simple task but the final code worked the first time. It illustrates the advantage of testing only portions of the code wherever practical. When testing unknown code on unknown hardware, minimizing the size of the unknown speeds debugging.

If using a monitor program or an emulator, the key debugging tool usually is the breakpoint. The code is run with a breakpoint set in the area where a problem is thought to be. When the breakpoint occurs, registers and memory are examined to determine the problem. The catch, in a real-time system, is that interrupts keep coming, motors keep turning, and in general, the real world keeps happening while the software engineer is trying to figure out what went wrong with the code. If regular edge-sensitive interrupts occur, all of them may be stacked up and waiting to execute when the code is resumed. While there is no silver bullet for this characteristic of real-time debugging, a few tricks can make this part of the task a bit easier:

- If the system will run this way, turn off all interrupts except those absolutely necessary to make the problem show up. If the system runs but the bug disappears, turn the interrupts back on one at a time until the problem comes back.
- If a timer interrupt is used, slow it down enough that you can actually see what is happening.
- If using debug trace outputs or a circular trace buffer, as described in Chapter 5, pay careful attention to the trace values when an error occurs, looking for patterns. You may find that a problem occurs only when a particular interrupt code appears in the table twice in a row, indicating a possible problem with reentrancy. Or you may find that a particular interrupt always occurs between two polling loop trace points, indicating a potential "shared memory or I/O trap" as described in Chapter 4.
- Ask yourself, "Did it work before the last software change?" I spent a long time one day asking a software engineer if she had changed the code since everything quit working. She kept telling me that she had not changed the code, then finally admitted to making changes, but insisted that the changes were not in the area that was not working. I finally convinced her to try the previous version and everything worked again. She had not realized that the new version required more stack space then she had allowed for.

- Determine whether the problem goes away when the emulator is connected. If so, this nearly always points to a hardware setup/hold time problem or a race condition—but not always. I once worked on a problem that disappeared every time the emulator was connected. The problem turned out to be the "shared memory or I/O trap" described in Chapter 4. For some reason, the emulator timing kept the interrupt from occurring between the two critical instructions.
- If you can, determine what specific condition causes the software to fail. This may be difficult without an emulator. If the exact hardware condition that caused the problem can be isolated, a pattern may emerge. Or a logic analyzer may shed light on the conditions surrounding the fault.

Debugging in RAM

Chapter 1 described the use of logic analyzers, emulators, and monitor programs for debugging. Use of debugging breakpoints requires that the code be executed from random access memory (RAM). A breakpoint replaces an instruction with different code that executes, essentially, a software interrupt.

Many emulators have internal RAM, and the external ROM is ignored when the internal RAM is enabled. The code is loaded into the emulator RAM and executed there.

Monitor programs must have RAM to set breakpoints. I have worked on systems where the system RAM was large enough to handle both the RAM requirements and the program code, so the code was just loaded into system RAM for debugging. In other cases, the PROM was replaced with RAM (the write signal from the processor must be wired to it) or a second RAM IC was added to the system.

ROM emulators contain RAM but plug into a PROM socket. They usually interface to a PC via a serial port and can be loaded from the PC with the target code. Breakpoints can be downloaded as well.

As processor speeds go up, emulators get harder to build. An emulator typically consists of some kind of pod that plugs into the processor socket, connected via a cable to additional emulator hardware, which then connects to a PC that controls everything. With processor clock rates exceeding 300 MHz, it is difficult to get anything to work over any kind of cable. In addition, modern high-integration processors often come in high-density surface mount packages. These parts do not use a socket, and there is no way to plug in an emulator. With leads on 0.03- or 0.02-inch or smaller centers, it is nearly impossible to clip anything over the package. A BGA or PGA package, with the leads underneath the IC, further complicates addition of any kind of emulator.

The bottom line to all this is that, with increasing processor complexity, development tools have regressed from emulators back to software debuggers. Fortunately, some technology improvements make debugging easier.

Many processors include on-chip debugging additions. Intel includes debugging registers on x86 processors from the 386 up. Four registers allow breakpoints to be set for reads, writes, or any (read/write) access to a specific location. The registers also allow 8-, 16-, or 32-bit memory locations to be monitored. So your debugging software can get some of the features of an emulator without additional hardware, even if your application is running out of ROM.

Motorola also supports on-chip debugging on many of its processors, including the 68K/Coldfire family and the 68HCxx microcontrollers. Motorola uses a technique called *background debugging mode* (BDM). BDM uses a connection to an external controller (such as a PC) to set breakpoints and examine trace information. One advantage of the Motorola scheme is that the processor can continue running while memory and registers are examined. This is a useful feature when the system is controlling motors or other devices that can cause damage if the processor stops.

As mentioned in Chapter 3, small microcontrollers often present a special challenge. In these devices, the processor, RAM, and program memory are self-contained, leaving the pins free for I/O functions. Consequently, there is no way to see what addresses are being executed and what RAM locations are being accessed. Sometimes you can build a special version of the controller that executes from external memory. But if your design uses all the controller's I/O ports, that is not an option. Many microcontrollers do not even support external memory. Accessing 64 K of external memory requires at least 19 data, address, and control pins, assuming data is multiplexed with addresses. Microcontrollers such as the Atmel AT90S2313 and many of the Microchip PIC devices come in 18- or 20- or 28-pin packages and simply lack enough pins to attach external memory. In a small microcontroller design using one of these parts, you have five disadvantages:

- You cannot use external RAM.
- You cannot debug out of RAM (you have to reprogram parts every time there is a change).
- You cannot set internal breakpoints (code is in PROM or flash memory).
- Often, not enough code space is left for a debugger, even if one were available.
- There is no way to see what internal addresses are being executed.

In a case like this, you are limited to using an emulator or adding debug techniques like those described in Chapter 5. Fortunately, the code for most of these designs is limited in complexity, if only because of limited code space, so debugging is less of a problem than for more complex processors.

Functional Test Plan

Once the software and hardware are thought to be working, a test should be performed to verify everything. A minimum test plan should describe every function to be tested. A minimal test plan for the pool timer might look like this:

Verify timekeeping.

Verify keypad operation and sensing.

Verify time rollover and on/off switching.

Verify override operation.

Verify water low sensor operation.

Verify time set.

Fortunately, if you have a good requirements specification, you also have a good start on the functional test requirements. The first thing you have to do in functional testing is make sure all the requirements are met. Some of these are a little difficult to prove—the "minimum switches and knobs" requirement for the pool timer is a good example—but the list of requirements is a good starting point for verifying the design. You can get software that tracks requirements in a document and you can carry the tracking through the test procedure to verify that the requirement was satisfied.

The problem inherent in any kind of minimal test plan like this is that it often omits several important steps. When you verify timekeeping, do you verify only that the time decrements or that it decrements at the right rate, rolls over at the right time, and check at least two cycles to be sure nothing hangs up? A more comprehensive test plan might have the following tasks:

Verify timekeeping:

Verify that time decrements correctly.
Verify timekeeping rate (accuracy in a real product would probably be specified in the requirements).

Verify keypad operation and sensing:

Keys are sufficiently debounced.
All keys except SET are ignored in powerfail mode.

Verify time rollover and on/off switching:

ON rollover, switch to OFF time.
OFF rollover, switch to OFF time.
Verify at least three cycles.

Verify override operation:

ON override in OFF timing.
OFF override in ON timing.
ON override in ON timing.
OFF override in OFF timing.
Override terminate in all modes.
OFF time expiration while in ON and OFF override.
ON time expiration while in ON and OFF override.
Correct LED operation in all modes.
Roll over time from 24:00 to override exit in all modes.
Override timekeeping in all modes.

Verify water low sensor operation:

LED operation (blinking ON LED).
Motor does not start in normal or override.
Normal operation resumes when level is OK again.

Verify time set:

Time increment and rollover, OFF and ON.
Time saved correctly when either or both times changed.
Override disabled for time set.

A functional test plan should cover not only normal operation of the system but abnormal cases as well. What if the operator presses the START button while the system is shutting down after someone pressed STOP? If you build enough systems, someone, somewhere, will try it. Trust me.

Sometimes special software can aid in this testing. For example, the pool timer software was modified to run the time at 20 times the normal rate so that the testing could proceed faster. There is danger here though, depending on how much has to change. At some point, a test of some kind has to be done with the final software.

It often is difficult for the software engineer to come up with all the possible scenarios that should be tested. A brainstorming meeting with other members of the project team, especially the software engineers, usually will produce a number of special conditions to check. If you are the only person on the project, bring in engineers from another project and have a code walkthrough/test scenario meeting.

Problem Log

On most projects, the first round of functional testing shakes out some problems. I recommend a problem log for any project, but especially for a large project. The log should list any errors encountered, how they were uncovered (what conditions or functionality was being tested), who is responsible for fixing the problem, and when it was fixed. An electronic log is sufficient, but a paper log near the test machine makes jotting things down easier. Do not try to use a problem log from the first day of debug, when the basic design is being debugged. The number and frequency of problems will make keeping the log up to date a chore, and it will not get done. This is especially true if you are working on multiple projects simultaneously, which, of course, you always are.

A problem log might seem like overkill for a project with one or two engineers, but it keeps minor problems that were put on the back burner from being forgotten completely.

When a problem is fixed, there is a question: How much testing is needed to prove the fix and verify that the fix did not introduce new problems? In a high-reliability system, it may be necessary to rerun the entire suite of tests. Some other systems may require only regression testing to verify the area of code or hardware that was changed. It is easy to say that only limited regression testing is needed, but be careful. A code change that seems innocuous may have far-reaching effects, especially if it interacts with other processes.

Once everything is working, the final step is to test the complete system against the original system requirements. This seems obvious, but sometimes a design team concentrates on the trees and forgets about the forest. The process of verifying the design at this level varies with the product, the company, and the customer and is beyond the scope of this book. It may include system testing that is several levels removed from the embedded control system, and it may involve testing in the field.

Chapter 7 discusses embedded designs that use multiple processors.

Multiprocessor Systems 7

The preceding chapters focused on single-processor systems and subsystems. Many embedded designs, especially in industrial or commercial systems, use multiple processors. The design of multiprocessor systems allows computing power to be distributed among different processors for redundancy, speed, modularity, or to simplify coding.

This book is about embedded systems, so this chapter does not address processor arrays used for high-speed computation nor will it cover multiple processors used in redundant voting or backup schemes. This chapter focuses on systems where multiple processors are used for distributed control of real-time events.

Multiple processors might be used in a project for number of reasons. One reason is modularity. For example, a particular processor-based subsystem needs to be installed only if a particular option is installed.

Distributed processors may simplify coding. Instead of one huge, complex, difficult to debug and maintain piece of code, the software is broken into much more manageable independent functions on different CPUs.

Although it seems counterintuitive, distributed processors may decrease costs. The processing horsepower to operate two tasks independently on separate processors may be considerably less than if one processor has to do everything. For example, a system may have to simultaneously handle high-speed events that require little processing but need extremely fast response, such as motor shaft encoder interrupts, and message-level interrupts that require extensive processing and perhaps a real-time operating system (RTOS). A single processor solution must be fast enough that the overhead of the message-level functions does not affect performance of the high-speed events and the repetition rate of the fast events does not slow down the more processing-intensive message functions.

A multiprocessor solution to this problem might involve a single-chip microcontroller for the motor control and an embedded processor for the message level tasks. This can be less expensive than a single processor powerful enough to do both.

Figure 7.1 illustrates a simple multiprocessor system for some undefined control application. CPU 1 handles the operator interface display and keypad. It communicates with CPU 2, which communicates with a higher-level host. CPU 2 also talks to CPU 3, which controls real-time motors and monitors event sensors. All three CPUs have their own programmable read-only memory (PROM), random access memory (RAM), and input/output (I/O).

In a real application, the display could be a liquid crystal display (LCD), light-emitting diode (LED), electroluminescent (EL), cathode-ray tube (CRT), or other type of display. CPU 2 communicates with a host. In a sorting application (moving plastic sheet, logs, documents, or anything else down a transport), CPU 2 might transmit the particulars about each item to a controlling host (data read from a document, size of a log, etc.) and receive decisions from the host as to what should be done with each item. CPU 3 might control motors that transport the items, or motors on an XY table, or the positioning of a tool in a NC (numerically controlled) machine tool. The event sensors could be position sensors that determine where the items are, pressure sensors, safety interlocks, or almost any event that the system needs to monitor.

The communication link between CPUs depends on the application. Communication between CPU 1 and CPU 2 could be an RS-232 connection since keypad/display data rates usually are not high. Communication between CPU 2 and the host could be anything from a slow RS-232 link to an Ethernet, SCSI, or Firewire connection, depending on the data throughput require-

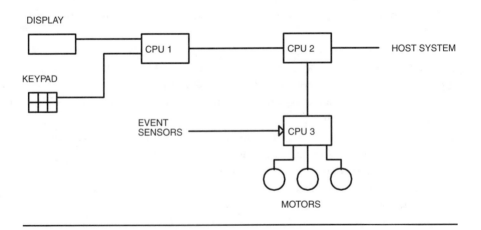

Figure 7.1
Simple Multiprocessor System.

ments. Communication between CPU 2 and CPU 3 could be a direct connection if both are on the same board.

Deciding how many processor subsystems to use and how to distribute the tasks among them is usually based on three considerations:

- Interdependence or modularity of the software.
- Processor throughput.
- Physical location.

Interdependent functions, such as positioning something using a stepper motor and reading feedback from the position sensor, are well suited to sharing one processor. In general, whatever arrangement minimizes the interprocessor communication usually is a good distribution of functions. In the previous hypothetical example, it makes no sense for CPU 2 to handle communication from the host and CPU 3 to handle communication to the host, unless the two information streams are independent and unrelated.

Code complexity is an issue as well. If CPU 3 is eliminated from the example and CPU 2 does both jobs, the software could become complex and difficult to develop. For example, CPU 2 might need to work at a message level to talk to the host, but getting continuous interrupts from sensors and motors makes the interactions hard to predict. While CPU 2 might be able to handle the *average* processing load, momentary peaks in taking care of motors might cause host data to be missed or serviced late.

Physical location may determine the breakdown of tasks. CPU 3 may be located remotely, near the motors. Even though CPU 2 is primarily a communication, as opposed to device control, processor, it might handle some minor sensor or I/O device that is located nearby.

In a system that has optional configurations, a processor might be dedicated to each option. The intelligence and cost to control the option (at least the low-level control) goes with the option.

Communication Between Processors

For multiprocessor systems where two or more processors are on the same board, several methods of interprocessor communication are possible.

Communication Register

Figure 7.2 shows probably the simplest mechanism communication between two processors on the same board. Data is clocked into an 8-bit register with tristate outputs, such as a 74AC374 (or, of course, part of a PLD), using a

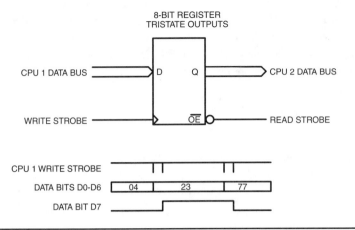

Figure 7.2
Register-Based Multiprocessor Communication.

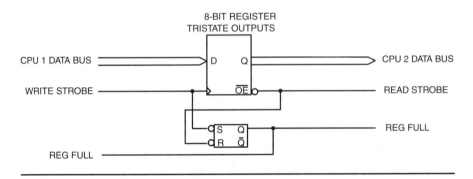

Figure 7.3
Register-Based Communication with Status Flip-Flop.

decoded write strobe from CPU 1. CPU 2 reads the data with a decoded read strobe. In this scheme, the lower 7 bits in the register (D0–D6) are used to transfer data and D7 is used as a strobe. Each time that CPU 1 wants to send data to CPU 2, it writes the data to the register, toggling the strobe line.

You will see in the timing diagram in Figure 7.2 that there actually are two writes by CPU 1. This is because there is a very small timing window where CPU 1 could be writing the register while CPU 2 is simultaneously reading the register. CPU 2 can then get bad data while the bits are changing. To prevent this, CPU 1 performs two write operations: the first one to change the data bits (D0–D6) and the second one to toggle the strobe. In the example shown in Figure 7.2, CPU 1 wants to send 23 then 77 to CPU 2. CPU 1 actually writes 23 then A3 then E7 then 77. By performing two writes per byte transferred, the low-order 7 bits that contain the data are guaranteed to be stable when CPU 2 detects that new data has been written. CPU 2 just polls the register, processing the new data every time the strobe bit changes state.

The drawback to this method is speed; there is no feedback to tell CPU 1 that CPU 2 has read the data. CPU 1 cannot send data any faster than the *slowest* speed at which CPU 2 polls the register. Say that CPU 2 is using a polling loop and has interrupts. In that case, the fastest CPU 1 can send data is the sum of the longest path through the polling loop plus the maximum interrupt stackup latency.

Figure 7.3 shows an addition to the circuit that can speed up communication considerably. This method still uses an 8-bit register with tristate outputs, but now a flip-flop has been added to the circuit. When CPU 1 writes the register, it sets the flip-flop; and when CPU 2 reads the register, it clears the flip-flop. So CPU 1 writes a byte and monitors the output of the flip-flop (which indicates register full). As long as the register remains full, CPU 1 cannot write another byte. CPU 2, on the other hand, monitors the register full output; and when the register is full, CPU 2 reads the byte, clearing the register full bit and enabling CPU 1 to write another byte.

The register full bit can be monitored by both CPUs using any input port bit, including a tristate status buffer, a port bit if one of the CPUs is a microcontroller, or a port bit on an I/O expander integrated circuit (IC).

This method speeds up the overall transfer rate, since CPU 1 can send data any time the register is empty. Note that the *slowest* transfer rate is the same as for the simple register/strobe arrangement. This is because the longest time CPU 2 may take to read the register is unchanged. However, since the *average* polling rate usually is faster than the slowest possible rate, the average throughput will be higher.

To speed things up even more, the register full bit also can be connected to an interrupt input to either or both CPUs. In this case, CPU 2 gets an interrupt when the register is full (or when the register *goes* full if the interrupt is edge sensitive) and CPU 1 gets an interrupt when the register is or goes empty. Now the average transfer rate can be quite high. The slowest rate is the sum of the worst-case interrupt latencies of both processors. However, both processors have to service one interrupt per byte transferred. CPU 1 does not know what CPU 2 is doing and may flood it with data at an inopportune time. The software for either CPU may need to disable the interrupt when performing time-critical processing. If this is necessary, the decrease in worst-case transfer rate needs to be taken into account.

If you are using processors with built-in DMA, you can use this technique to implement a very fast communication scheme. The register full output is connected to the DMA request of CPU 2 and the inversion (register empty) is connected to the DMA request of CPU 1. CPU 1 puts the data it needs to send into a block of memory and programs the DMA controller to send it. CPU 2 programs its DMA controller to read data from the register and put it in memory. Now the two DMA controllers handle the transfer, typically at very high rates.

The problem with DMA-controlled transfers is this: How does CPU 2 know how many bytes to transfer? There are three solutions to this problem:

- The first DMA technique is very simple. All transfers are a specified size, such as 256 bytes. If the data to be transferred are shorter than that, it is padded out (with zeros or some other constant value) to the block size.
- The second technique involves a length byte. The first byte transferred by CPU 1 is a length value. CPU 2 sets up its DMA controller to transfer 1 byte and generate an interrupt when done. When the length byte is received, CPU 2 services the interrupt and sets up the DMA controller to receive the specified number of bytes. This method requires CPU 2 to service two interrupts for every transfer.
- The third technique requires a second interrupt path between the two processors. CPU 2 sets up its DMA controller to transfer more than the *maximum* number of bytes in an actual message. If the longest message is 64 bytes, then CPU 2 sets up the DMA controller to transfer any value greater than 64 bytes. CPU 1 sets up the DMA transfer and, when it is completed, notifies CPU 2 via the separate interrupt path. CPU 2 reads the number of bytes transferred from its DMA controller and then processes the received data. Note that the CPU 2 DMA controller will never generate an interrupt, since it never transfers the number of bytes programmed.

You can use the DMA technique even if only one processor supports DMA. The non-DMA processor can poll the register to see when data are available. The speed is no higher than a polled register approach, but whichever processor has DMA is relieved of the need to poll for each byte.

Figure 7.4 illustrates a variation of this DMA method. This was designed for a dual-80188 application where CPU 1 had both DMA channels used for something else. The DMA channels of CPU 2 were used for data transfer.

This scheme uses two 8-bit registers for bidirectional communication. Register 1 transfers data from CPU 1 to CPU 2 and Register 2 transfers data from CPU 2 to CPU 1. The register full bit for Register 1 drives DMA channel 0 on CPU 2 and the register empty bit on Register 2 drives DMA channel 1 on CPU 2.

In addition, there is one interrupt from CPU 1 to CPU 2 and one from CPU 2 to CPU 1. The CPU 1 to CPU 2 interrupt is set by CPU 1 and cleared by CPU 2. The other interrupt is set by CPU 2 and cleared by CPU 1. Both

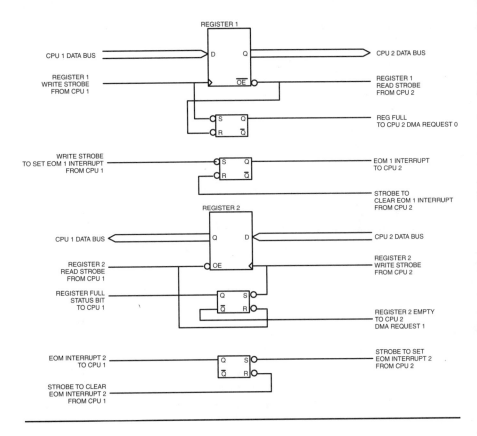

Figure 7.4
Dual 80188 Communication Using a Single CPU DMA.

interrupts are available as status bits to both CPUs. Interrupt set and clear, of course, are decoded read/write strobes.

The sequence of events for transferring data from the CPU1 to CPU2 is as follows:

CPU 2 sets up DMA channel 0 to transfer 256 bytes (or any value greater than the longest possible message).

CPU 1 sends the message. CPU 1 must allow sufficient time between successive bytes to permit the DMA transfer to complete.

CPU 1 sets the interrupt after the last byte is transferred.

CPU 2 services interrupt. This includes terminating the DMA transfer, reading the DMA controller to determine how many bytes were transferred, setting up the DMA to receive the next package, and determining if the message must be processed immediately or if it can wait. After interrupt processing is complete and *the DMA is set up for the next transfer*, CPU 2 resets the interrupt.

When the interrupt goes inactive, CPU 1 can send the next message.

The sequence of events for transferring data from CPU 2 to CPU 1 is as follows:

CPU 2 sets up DMA channel 1 to transfer the message from memory to the communication register. DMA is set up to interrupt CPU 2 at end of message transmission.

CPU 1 reads each byte as it is available in the communication register.

When complete message is transferred, the DMA controller interrupts CPU 2, which then sets interrupt to CPU 1.

When CPU 1 clears the interrupt, CPU 2 can send the next message.

The only possible problem here is that CPU 1 must not transfer data too fast to CPU 2. One way to prevent this is to have CPU 1 poll the register 1 full bit and not transfer if the register is full. However, if CPU 2 is not performing operations that prevent the DMA from acquiring the bus or is not considerably slower than CPU 1, a minimal software delay should be adequate.

A problem can occur with any of the register and flip-flop methods if either CPU is considerably faster than the other, such as if one is a digital signal processor (DSP) and the other is a relatively slow microcontroller. If CPU 1 is faster than CPU 2, then CPU 1 may detect the register full going inactive and write a new byte while CPU 2 still has its read strobe active to read the first byte. If CPU 2 is faster, it may detect the register full condition and read the byte while CPU 1 still has the write strobe active. In either case, the SR flop will end up in the wrong state, causing a byte to be missed or read twice.

Two solutions to this problem are to add a delay between register full/empty detection and the next read or write for the faster CPU. Another solution is to use a "D" type register full flip-flop with both asynchronous set/reset and a clock input. The slower CPU drives the clock to set or clear the flip-flop. This ensures that the flip-flop is set or cleared (depending on which CPU is slower) at the *end* of the read or write cycle.

Figure 7.5 shows this problem. In Figure 7.5A, CPU2 is much faster than CPU 1 and polls the data flip-flop twice during the CPU 1 write. Consequently, CPU 2 thinks two bytes have been written instead of one. Note that since the data actually are not written to the data register until the end of the write cycle, the first byte that CPU 2 reads is the *previous* byte that was written. The

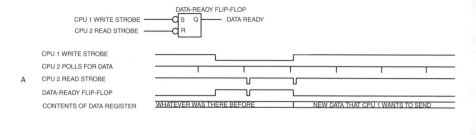

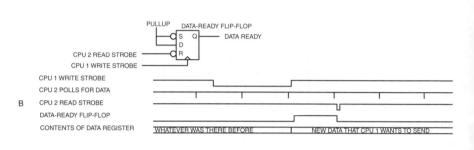

Figure 7.5
Fast/Slow CPU Communication Timing Problem.

diagram shows the data-ready flip-flop going low during the CPU2 read cycles, although real hardware may or may not do that, depending on what type of ready flip-flop used.

Figure 7.5B shows how using a D type register, such as a 74ACT74, fixes this problem. Now the data ready goes low only after the *end* of the CPU 1 write cycle, and everything works as it should.

Of course, for two-way communication, these methods can be expanded by adding another communication register, written by CPU 2 and read by CPU 1. Wider registers can be used with 16- or 32-bit processors. You can mix techniques as well. Say the CPU 1 to CPU 2 path requires a lot of data at high speed, but what comes back from CPU 2 to CPU 1 is infrequent single-byte status responses. In this case you might use a DMA scheme to send data from CPU 1 to CPU 2 and a polled register and flip-flop for the reverse path.

The communication protocol for using a register of this type depends on the data that must be exchanged. If CPU 2 just gets simple commands like "Turn on motor 1" and "Turn off motor 2," each command can be a single byte or even a bit in a byte. If the commands need to be more complex, a string of bytes can be used where the first byte is an opcode that determines what the operation is and how much data follows. One opcode, for example, might be "Move up the NC head" with one or more subsequent bytes to determine how far the movement should be.

In cases where the data length varies, the first byte can state the length, or the first byte can be an opcode and the second byte state the length. For any multibyte protocol, a checksum byte can be added to detect errors or missed bytes.

FIFO Devices

A second method for interprocessor communication involves FIFO (first in, first out) buffers. Conceptually, this is the same as the register approach, except that FIFO buffers replace the register, with one CPU writing the FIFO buffer and the other CPU reading the FIFO buffer. The FIFO buffer holds the data, allowing CPU 2 to read the data in order at its convenience. Most FIFO buffers have a pin that tells when the FIFO buffer is empty. This can be monitored to determine when data are in the FIFO buffer (not empty = data available).

In addition to cost, the drawback to using FIFO buffers is message stackup. In some systems, messages have different priorities. One message might be information to write to an LCD display, something that can be handled when the processor gets to it. A subsequent message might be a command to perform an emergency shutdown or some other task that cannot be delayed. In a situation where everything has low priority, CPU 2 can leave messages in the FIFO until it can get to them. If high-priority messages can follow low-priority messages, CPU 2 has to take the time to read each message immediately on the assumption that it or the following message has a high priority. If the message has low priority, CPU 2 has to buffer the information in RAM and process it later. This results in two drawbacks. First, low-priority messages must be handled twice: once to read and store and again to process. The second drawback is that even low-priority messages must be taken from the FIFO immediately, because a high-priority message might follow. This forces every message to be treated as a high-priority message.

If the register and flip-flop approach with DMA is used, the message stackup problem can be reduced. An advantage of the DMA method is the ability to set priorities for message servicing. The receiving processor can check the opcode of each message after it is received. If the message has low priority, it can be left in memory for later processing. Of course, this method requires that the receiving processor set up multiple DMA receive buffers to hold several messages and keep track of which messages have been serviced.

Dual-Port RAM (DPRAM)

In cases where a lot of data must be transferred between two processors, a dual-port RAM is common. Dual-port RAM is shared between two processors. If both processors want to access the RAM at the same time, one has to wait until the other is finished.

Some dual-port RAM ICs handle arbitration internally. These devices have a signal to each processor to request a wait state for arbitration, or they use a *synchronous* dual-port memory architecture that permits simultaneous access by both processors. The 70908 from Integrated Device Technologies (IDT) is a 64K × 8 dual-port RAM. The IDT 7052 is a 2K × 8, four-port device that can allow four processors to communicate using a common RAM area.

One drawback to using synchronous dual-port RAM ICs is possible data corruption. If one processor writes to a location while the other is reading, the write may not be completed correctly or the read data may be corrupted.

For cost-sensitive designs, an inexpensive way to produce a dual-port RAM is to use the bus hold capability of one CPU.

The block diagram in Figure 7.6 illustrates a means to implement the hold-based dual-port RAM. This example uses two Intel-style processors, such as

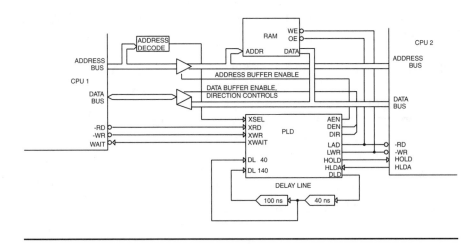

Figure 7.6
Dual-Port RAM Using a Bus Hold.

the 80188, but the concept can be adapted to any processor that has bus-hold capability.

CPU 1 has an address decoder that selects its local RAM, ROM, I/O, and access to the other processor. For simplicity, the CPU 1 RAM, ROM, and I/O are not shown.

A memory map for CPU 1 might look like the following:

00000 to 1FFFF	128 K local RAM
20000 to 27FFF	32 K dual port RAM
F0000 to FFFFF	64 K EPROM

When CPU 1 accesses locations 20000–27FFF, the address decoder generates a select signal (XSEL) to the programmable logic device (PLD) in Figure 7.6. The read and write strobes are generated as well. The PLD responds with a wait request to CPU 1 and asserts HOLD to CPU 2. CPU 2 releases its bus, tristating the address, data, and control lines. When CPU 2 releases its bus, it responds with HLDA. The PLD then enables the address and data buffers that connect the CPU 1 bus to the CPU 2 bus. After a setup delay, the PLD asserts the read or write strobe to the RAM and removes the WAIT request to CPU 1. During all this, CPU 2 is in a hold state and does not drive the bus. When CPU 1 completes the read or write cycle, the PLD tristates the read and write lines and removes HOLD. CPU 2 then removes HLDA and reacquires its local bus. This example uses a delay line for illustrative purposes; the same thing could be accomplished with a synchronous design.

The CPU 2 address decoding logic is not shown on Figure 7.6. This logic must recognize accesses from both CPUs and generate the RAM -CE signal. Note that the RAM does not need to be in the same place in the CPU 2 address space as it is in the CPU 1 space.

The PLD equations for this are as follows:

```
// PIN DESCRIPTIONS
// !XWR,!XRD: EXTERNAL PROCESSOR READ AND WRITE STROBES.
// !XSEL: ADDR DECODE FROM EXTERNAL PROCESSOR.
// HLDA: FROM LOCAL 188.
// DL140: 140 NS OUTPUT OF DELAY LINE
// DL40: 40 NS OUTPUT OF DELAY LINE
// DIR: DIRECTION CONTROL FOR ACT245 BUS BUFFER.
// 0 = READ FROM LOCAL BUS TO EXERNAL BUS.
// !DEN: ENABLE FOR ACT245 BUS BUFFER
// !AEN: ENABLES ADDRESS BUFFERS FROM EXTERNAL ADDR BUS
// TO LOCAL ADDR BUS.
// !LRD: READ STROBE TO LOCAL BUS
```

```
// !LWR: WRITE STROBE TO LOCAL BUS
// HOLD: TO LOCAL 188.
// !XWAIT: TO EXTERNAL CPU
// DLD: DELAY LINE DRIVE
// FF1, 2: MEMORY LATCH


// THIS PLD WILL ARBITRATE THE LOCAL BUS FOR EXTERNAL
   ACCESS.
// REQUESTS HOLD FROM 188, THEN ALLOW EXTERNAL BUS
   ACCESS
// WHEN HLDA IS RETURNED. TO PREVENT PROBLEMS WITH
// BACK-TO-BACK CYCLES. A SECOND ACCESS WILL NOT BE
// PERMITTED UNTIL THE HOLD ACKNOWLEDGE FROM THE FIRST
   ACCESS
// HAS BEEN REMOVED BY THE LOCAL 188.


XWAIT = XSEL & !FF2


HOLD = XWAIT & XRD & !HLDA
     # XWAIT & XWR & !HLDA
     # HOLD & XRD
     # HOLD & XWR


// AFTER HOLD AND HLDA BOTH TRUE, TIMING CYCLE STARTS.
// EXTERNAL ADDRESS IS ENABLED FIRST. AFTER 40 NS SETUP,
// LOCAL READ/WRITE IS ASSERTED. 100 NS AFTER THAT,
// XWAIT IS REMOVED TO COMPLETE CYCLE.


DLD = HOLD & HLDA & XRD & !FF1
    # HOLD & HLDA & XWR & !FF1
FF1 = DL140
    # FF1 & !FF2


AEN = HLDA & HOLD & !FF2
    # AEN & LWR
    # AEN & LRD


DEN = AEN
DIR = XRD # LRD


FF2 = FF1 & !DLD & !DL140
    # FF2 & HOLD
```

```
LWR.OE = DEN
LRD.OE = DEN

LWR = XWR & DL40    // FOLLOWS BUS WRITE
    # LWR & DEN & XWR

LRD = XRD & DL40    // FOLLOWS BUS READ
    # LRD & DEN & XRD
```

The one drawback to using this dual-port RAM technique is that both processors are slowed down by the access. A dual-port RAM IC or controller IC will place one processor in a wait state only if both attempt simultaneous access. In this design, CPU 1 must wait while CPU 2 gets into a hold state, so excessive access by CPU 1 can affect throughput of both processors. However, this can be a cost-effective design, since the dual-port RAM can be the CPU 2 local RAM.

Transferring data between processors in a dual-port RAM can be accomplished in a number of ways. One method is to have one or more sequential buffers with semaphores. For example, RAM locations 1000–1100 (hex) might be configured into four buffers as follows:

1000:	Semaphore buffer 1
1001–103F:	Buffer 1, 63 bytes
1040:	Semaphore, buffer 2
1041–107F:	Buffer 2, 63 bytes

. . . and so on through buffer 4.

In operation, CPU 1 puts data in buffer 1 then sets semaphore 1. CPU 2 sees semaphore 1 set, processes the data, and clears semaphore 1.

The next block of data from CPU 1 goes in buffer 2, then buffer 3, then buffer 4, and then back to buffer 1. If CPU 1 wants to put data in a particular buffer and the semaphore still is set, the buffer is not available and CPU 1 must wait.

If the messages have variable length, the semaphore may be replaced with a length byte (or word). CPU 1 places data in the buffer then places the length at the first byte. CPU 2 clears the length to 0 when it has processed the data. This makes more efficient use of the RAM, since the buffer length is only as long as needed for a particular message and subsequent messages can be strung together in memory, the length byte of one message immediately following the last byte of the previous message. However, it makes the code less efficient, since the CPU has to search through the buffers using the lengths to find the first unused one.

The length/semaphore must be set by the sending CPU only after the complete message is in the buffer, or the receiving CPU may see the length byte and try to read the message before it is completely written.

Data corruption in synchronous dual-port RAMs already has been mentioned. Any type of dual-port RAM arrangement is susceptible to data corruption if the memory is managed poorly. In general, data buffers should be segregated into send and receive buffers. One CPU writes to the send buffers, while the other CPU reads them, and the reverse is true for the other set of buffers. This arrangement is needed because, if the buffers are shared, both processors may try to simultaneously grab an empty buffer.

If it is impossible to segregate the buffers this way, then a protocol must be put in place to keep both processors from attempting to access the same buffers at the same time.

An additional problem can occur with when using 8-bit dual-port RAM with 16-bit processors. If the semaphores and buffers are 16-bit words, the processors will have to do two 8-bit memory cycles to access a 16-bit semaphore. It is possible for one processor to access a memory location in the middle of the two write cycles from the other processor.

This problem can be avoided if the processors have a LOCK function, which can be used to lock out access to the DPRAM by the other processor. However, this will not work with a synchronous DPRAM design. In general, it is safest to have critical semaphores be 8 bits wide in these applications. Use 8-bit semaphores to control access to buffers, and if necessary, use 16-bit counters and data values.

Serial Communication

Chapter 4 describes a method of communicating between a pair of ADSP-2101 processors using the built-in synchronous serial port. In that example, the serial interface sent 16 bits at a time. The low-order byte (D0–D7) was designated as data, D8 and D9 indicated the source of the transmission (up to four DSPs were possible in the system), D10 and D11 indicated the destination, and D12–D15 were an opcode that indicated what the data were for. While this scheme required 16 bits to be transmitted per byte, most opcodes require only 1 byte, and the mechanism allows multiple devices to share the bus.

The Microwire and I²C buses described in Chapter 2 can be used for interprocessor communication, albeit somewhat slowly. In this scheme, one processor typically controls the bus as a master, and the other responds like a peripheral device. However, the I²C specification supports multimaster operation in a fairly unique way.

The problem with any shared multimaster bus is arbitration—which master gets the bus when two or more want it at the same time. Some arbitration schemes allow multiple masters to transmit, detect the bit errors, and resend the bad transmission. I²C performs arbitration by allowing any master to send when the bus is idle. If two masters attempt to send at the same time, eventually a bit will be in the data stream where one master is sending a 1 and the other is sending a 0. Since the I²C bus is an open-collector device, the master sending the 0 will pull the data wire low. At this point, the other master is expected to sense that the state of the bus is different from what it is sending and turn off its drivers.

There are a couple of interesting things about this arbitration method: First, it does not cost any time—transmissions proceed as they would in a single-master system. Second, no priority is assigned to the bus masters. Which master wins control of the bus is completely dependent on the data each is transmitting. Third, the point in a transmission where arbitration is decided is completely data dependent. If two masters attempt to send information to the same address, control of the bus may not be decided until well into the transmitted data fields. Figure 7.7 illustrates I²C bus arbitration between two masters. The 3.4 Mbits/sec high-speed mode of I²C does not support multimaster transmissions.

Sending I²C over long distances is somewhat problematic: Both the SCL and SDA lines are bidirectional and open collector, so they cannot just be buffered with RS-485 buffers unless the bus is completely implemented in software. The usual method of buffering an I²C bus involves a circuit that senses current flow to determine if a device is trying to drive the line and turns on the right buffer to send data the right direction.

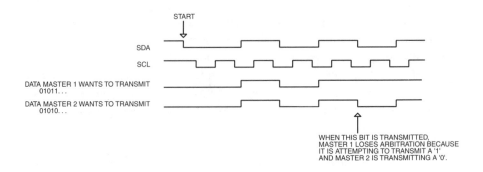

Figure 7.7
I²C Bus Arbitration.

Some microcontrollers, such as the 8051, have synchronous serial ports that are suitable for interprocessor communication. These generally run at a fairly high submultiple of the processor clock for fast data transfer. Since they usually are half duplex, a handoff protocol must be established for bidirectional communication.

Processors on Different Boards

In systems where two processors on separate boards need to communicate, several methods are available. A serial RS-232 link has already been discussed.

Some port expander ICs, such as the Z8536, have a built-in communication mode where 8 data bits of one port and 2 or 4 bits of another port can be interconnected between two devices to make a bytewide interface with interlocked handshake. If the communication distance warrants, the interface can be made differential or otherwise noise immune.

An asynchronous serial interface can be used without the RS-232 interface. Some high-speed UARTs can operate up to 1 Mbit per second. If an RS-485 differential interface is used as illustrated in Figure 7.8, several processors can be connected in a high-speed party line arrangement. Note that the RS-485 party line communication bus can be quite long and interconnect subsystems over significant distances. Although not covered in detail here, more complex communication schemes involve Ethernet, Firewire, or other standard, high-speed interfaces.

CAN Bus

The CAN (controller area network) is a serial bus originally developed for use in motor vehicles. It is a multimaster bus that supports multiple, equal nodes. The nodes have no specific address. Address information is contained in the identifiers of the transmitted messages. Nodes may be plugged in and removed while the system is operating ("hot swapping").

The CAN bus is a 120-ohm differential serial party-line bus. Three bus speed ranges are available: Low speed (ISO-IS 11519-2) defines a Class A bus with speeds up to 10 kbps, and a Class B bus for speeds from 10 kbps to 125 kbps. The high-speed specification (ISO-IS 11898) defines a bus with speeds between 125 kbps and 1 Mbps.

CAN uses NRZ (nonreturn to zero) signaling, with bit-stuffing to allow resynchronization. The CAN differential lines have two states: in one state, both lines are driven to 2.5 V and in the other state one line is driven to 1.2 V and the other to 3.5 V. This gives a differential voltage swing between 0 and 2 V.

CAN messages consist of a start-of-frame bit, followed by an arbitration field consisting of 12 bits: The 11-bit identifier, which reflects the contents

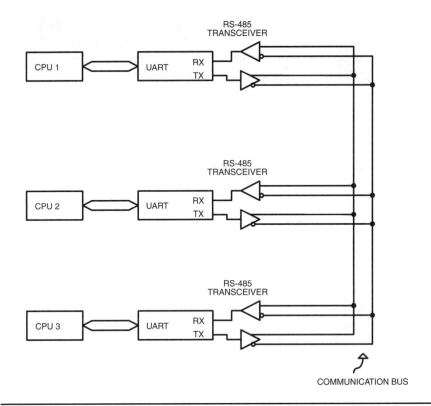

Figure 7.8
RS-485 Party Line Multiprocessor Communication.

and priority of the message, and the remote transmission request bit. The arbitration field is used to arbitrate between transmitters. If multiple transmitters attempt to gain control of the bus at the same time, the nodes with lower-priority messages will drop out during the arbitration field, leaving the node with the highest-priority message in control of the bus.

Next is the control field, consisting of 6 bits. The first bit of this field is called the IDE (identifier extension) bit, and the next bit is reserved. The remaining 4 bits are the data length code (DLC) and specify the number of bytes of data contained in the message (0–8 bytes).

The data follows the control field and consists of however many bytes were defined by the DLC. After the data is a 15-bit CRC (cyclic redundancy check) for error checking. Following the CRC is an acknowledge field, where the receiving node drives an acknowledge bit onto the bus to notify the transmitter that the message was correctly received. Last, 7 empty bits complete the frame.

CAN error checking is performed by three methods: First, the CRC (a complex checksum method) is calculated and inserted into the message by the transmitter. The receiving node calculates the same CRC and compares it against the received CRC to detect transmission errors. If a CRC error is detected, an error frame is generated to request retransmission.

The second error check uses the acknowledge bit; the message is sent from the transmitter to the receiver, but the acknowledge bit is sent from the receiver to the transmitter. If no acknowledge bit is received, the message is retransmitted.

Finally, a frame check is performed by the transmitter, by looking for an incorrect state during the CRC delimiter, acknowledge delimiter, end of frame and interframe space periods. An incorrect signaling value during these periods is an error.

There are two versions of CAN: version 2.0A (Standard CAN) supports an 11-bit identifier field (supports 2047 message types) and version 2.0B (Extended CAN) supports an 18-bit identifier extension, for a total 29-bit identifier field.

CAN interconnects can be up to 40 m long at 1 Mbps. Longer cables can be used with lower bit rates. Up to 30 nodes may be connected to a single CAN bus. A number of manufacturers make microcontrollers that interface directly to CAN bus. Examples are the Siemens C167R and Intel 87C196CB. Intel also makes a communications controller, the 82527, that provides a CAN interface for processors that lack embedded CAN capability. Figure 7.9 shows the data sequence and voltage levels for CAN.

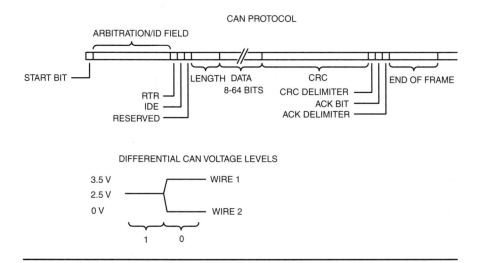

Figure 7.9
CAN Bus.

Open-Collector Serial Interface

Figure 7.10 shows a simple means to provide interprocessor communication using an asynchronous serial port such as the one available on most microcontrollers. All the processors drive a common serial line with open-collector

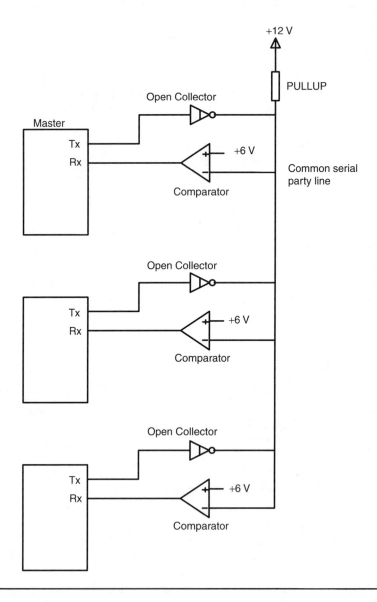

Figure 7.10
Serial, Asynchronous Communication.

drivers. The common serial line is pulled up to +12V. Each processor has a comparator, referenced at +6V, to receive data.

With a 6V reference, the noise immunity of this approach is similar to that of RS-232, but the open-collector drive allows multiple devices to communicate over a single wire. Since the system uses standard asynchronous signaling, any type of processor can communicate on the bus.

To implement this system, one of the processors normally would be designated as the master, and the other processors would transmit only when requested to do so by the master. This avoids bus contention.

Figure 7.11 shows a variation on the open-collector serial communication method that allows a slave to request attention from the master. To implement this, the common serial line is pulled to +24V instead of +12V. The master has two comparators, one referenced at 6V for the data, and another comparator referenced at +18V and driving an interrupt on the processor.

The slaves can request attention by pulling the common serial line down with a 12V zener diode. When no slave is requesting attention, the common line swings between 0 and 24V. When a slave is requesting attention, the serial line swings between 0 and 12V. So the master can monitor the request input when the serial line is idle to determine if any slaves are requesting attention. The slave devices must be polled by the master to determine which ones need service.

The maximum baud rate for this method usually will be lower than for the +12-V-only system. At +12V, a 600-ohm resistor dissipates about 0.25W. But at 24V, a 2300-ohm resistor dissipates the same power. So the 24V system typically will use a larger pull-up, resulting in a lower maximum data rate. But this communication method allows multiple processors to communicate, with an attention request capability, over a single wire (plus ground).

Parallel Port Interface

Many single-board computers, such as PC/104 systems (see Chapter 9) include a parallel printer port, compatible with that found in the IBM PC-clone world. In many embedded systems, this port is not needed to communicate with a printer. The standard printer port provides eight data lines, a strobe signal, four output lines, and six input lines.

Two computer boards can be interconnected using their printer ports. There is a standard interface for this, called Interlink, used to interconnect PCs. Off-the-shelf software and cables are available to implement this interface.

Interlink communicates using only four of the data lines because the data lines on the standard printer port are unidirectional, output only. You cannot tie the data lines of two printer ports together or you will get bus contention. However, many modern printer ports support various bidirectional modes of

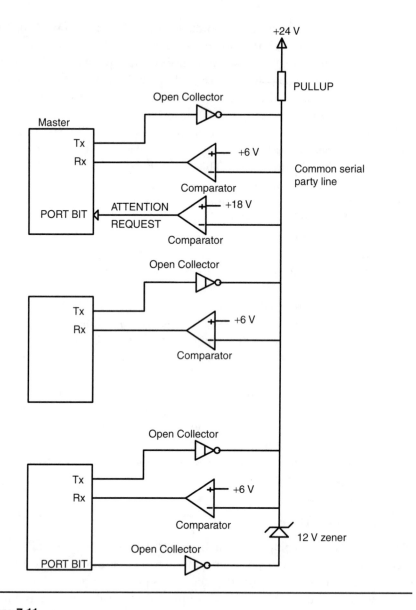

Figure 7.11
Serial, Asynchronous Communication with Attention Request Feature.

operation. You could use this capability to get full 8-bit-wide transfers between two computer boards, but you need some additional hardware to isolate them, since both boards will come up in standard mode, with the output drivers enabled.

Acknowledge Timing

In many multiprocessor systems, one higher-level controller passes commands to lower level controllers. These commands usually cause the lower-level controller to perform some action—a command to "plane the block smooth" would be an example in the moving block scenario.

One issue in any system of this type is how and when to acknowledge the command. There are basically four possibilities:

- **No acknowledge.** In this scheme, the low-level controller does not acknowledge the command at all. However, there may be an acknowledge that the data were taken, such as the register empty/full bit associated with the communication register. The higher-level controller has no indication of when or how the command was carried out.
- **Acknowledge on error.** The low-level controller sends an acknowledge indication only if there was an error in communication or in carrying out the command. For instance, if the command is to move a robotic arm to a certain position, an error would be returned if the arm is stuck.
- **Acknowledge on receipt of command.** The low-level controller acknowledges that it has received the command. If the communication protocol includes a checksum or other error check, the acknowledge will include an indication if there was an error. The higher-level controller still has no indication of when the command actually is executed.
- **Acknowledge on completion of command.** The low-level controller acknowledges when the command has been executed—when the arm has been moved, to use the robotic arm example. The higher-level controller now knows that the command was received and executed and when the command was complete. However, if the communication protocol does not allow multiple commands to be sent, then the higher-level controller is inhibited from sending additional commands until the previous command was executed and acknowledged.

These protocols can be combined. For instance, every command might be acknowledged when received, but an execution acknowledge is sent only if there is an error.

Any scheme that does not force the higher-level controller to wait for acknowledge of execution before sending additional commands must have a mechanism to handle errors. If an execution error occurs in command A, but commands B and C have already been sent, how does the higher-level controller know which command did not execute? What if command C depends on command A executing correctly? For example, our robotic arm might be told to move to a certain position (command A), insert a tool in a slot (B), and turn it (C). If the first command did not execute, then the last one is pointless and may even cause damage. So the protocol needs to specify what happens in case of an error. If commands can be "pipelined" (a new command sent before the old one is executed), you need to stipulate how many commands can be allowed to stack up so that the buffers do not overflow.

Design Pitfalls

Multiple Measurements Be careful of having two processors measure one thing. Because the "thing," whatever it is, will be measured with a digital system, there always is the possibility that the two processors will get different results. If they are measuring time, there will be at least one clock ambiguity in the measurement. If they are measuring a voltage, there always will be an ambiguity of one count in the two ADC output results. This can be a problem if there are fixed thresholds. For instance, if you are moving wooden blocks down a conveyor system and one processor determines that the length of a block is (just barely) too long, be sure another processor will not declare it to be OK. The first processor might skip sending the block to a planing process, while the second one proceeds with some other process that depends on a smooth surface. If there are fixed thresholds for what you measure (too short, too long, too heavy, voltage too high, etc.) be sure that the first processor that detects an error overrides the measurements of all subsequent processors. Or else be sure that a conflict does not cause problems later.

Synchronization Say you have a process controlled by multiple processors, like the wooden block example just mentioned. One processor cuts the blocks to size, the next one planes them smooth, the third one stamps a logo on the blocks, and so forth. Say that everything in the system is synchronized to a clock that occurs once each time that the conveyer system moves 0.1 inch. If data are passed between processors as each block moves between the regions controlled by each processor, there is a risk of a one-clock ambiguity in the timing. Be sure these cannot add up as the blocks move along. Either keep the time increment small enough that the cumulative error is no problem or resynchronize each processor to the leading edge of each block. This may require more sensors than otherwise would be required for system operation.

Revisions With a multiprocessor system, it often is possible to change the firmware for one processor without changing the others. Be sure this causes no problems if some function works differently than before. For instance, a new firmware revision might handle error messages from another processor with a different priority than the original firmware. Or the maximum buffer size might get changed in such a way that it is a problem only if certain errors occur. You may need additional regression testing of the combined system when firmware is changed.

It is not a bad idea to have a suite of tests that is run any time firmware changes are made to any of the processors in the system. This would need to test all the error conditions and all the communication paths, buffers, and types. Of course, this type of error can creep into a single-processor system as well, but it is easier to overlook in a multiprocessor system due to the isolation of the CPUs.

Error handling   Be sure all the processors handle errors consistently. In the block-of-wood example, if a problem occurs, do not let one processor try to stop everything while another tries to keep the conveyer going so everything falls off the end.

Berserk processors   Where possible, handle the case of a berserk processor that writes all through memory or a frozen processor that will not communicate at all. Have timeouts on communication operations. You usually cannot operate normally, but at least make all the moving/rotating mechanisms safe. In cases where you have optional subsystems, the rest of the system may need to operate normally when something in the optional part is not working.

Extreme isolation   In a multiprocessor design, it is tempting to isolate functions so that one processor handles all of one function, independent of the other processors in the system. This makes for a modular design. However, in a design where there is a chance that things might change, make provision for the master control CPU (if there is one) to alter parameters. In the wood block example, the planer might plane the blocks to a certain smoothness. However, once in production, it may be necessary to change that parameter. This might be because a new type of wood is encountered or because a sensor went out of production and the new sensor isn't quite identical.

In a case like this, it is a good idea to make the smoothness parameters (however they are measured) modifiable. You might have the system reset to the default parameters, but allow the master CPU to change them if necessary. Of course, it is difficult to predict what might change, but some effort in this area often pays off.

This approach is especially helpful in a system where the master controller is a PC with software that can be downloaded or upgraded via CD-ROM, while the lower-level controllers are PROM-based microcontrollers. For many companies, changing the microcontroller code means sending out a service engineer (expensive), while the host PC code might be upgraded just by sending the software to the customer.

For the same reasons, you may want to consider adding hardware that would allow the master controller to reprogram the lower-level processors. This implies the use of microcontrollers that are capable of in-circuit programming, of course.

Locking problems   Data corruption in dual-port RAM systems has already been mentioned, but we look at a specific example here. I got a call one day about a dual-processor system that had been designed by an outside design house. When the firmware was upgraded, an intermittent problem suddenly showed up.

It did not take long to determine that the problem was corruption in the dual-port RAM. One processor was attempting to perform a read-modify-write operation on a semaphore. Occasionally, the other processor would attempt to write to the semaphore, in between the read and write operations of the first processor. This corrupted the memory.

The processors had a lock output that indicated when the CPU was attempting an operation that could not be interrupted; the DPRAM controller was supposed to lock out the second CPU while the first was accessing the memory. However, a design flaw in the controller allowed the second CPU access to the memory even though the LOCK signal was supposed to prevent it. It was supposed to work like this:

CPU 1 reads the semaphore, asserting the LOCK signal.

CPU 2 requests access to memory and is put in a wait state due to the LOCK.

CPU 1 writes to the semaphore.

CPU 2 is released from the wait state, reads the semaphore, finds the value that CPU 1 wrote.

It actually did this:

CPU 1 reads the semaphore, asserting the LOCK signal.

CPU2 requests access to memory, gets memory, reads the semaphore, finds it is 0.

CPU 1 writes to the semaphore.

CPU 2 writes to the semaphore, overwriting the CPU 1 value.

The problem showed up when it did because the firmware change altered the relative timing of the two processors so that they occasionally conflicted in accessing the memory. Although this problem occurred because of a flaw in handling the LOCK signal, a similar circumstance can occur any time that two (or more) processors try to write to a single RAM location. The following are some guidelines for using multiport RAM:

- Wherever possible, do not have two processors that write to one memory location. As mentioned earlier in the chapter, segregate buffers so that one processor writes to a buffer and the other one reads.
- Never have a situation where two processors can simultaneously check a memory location (such as a semaphore) and then write to it. This is a sure way to get contention. Have one CPU write a flag location to indicate that data are in the buffer, have the other CPU write the location only when it has taken the data. Then the two CPUs do not contend for the location at the same time.
- If you must have a resource (such as a buffer) that is shared between multiple processors, have a two-step arbitration protocol. In this scheme, each CPU writes a unique code to a semaphore to indicate it wants the resource (whatever the resource is). Then each CPU checks the semaphore (preferably twice) to ensure that its own code is written there. If a conflict occurs, whichever CPU's code is left in the semaphore wins.

The last guideline works like this (in this example, a nonzero value in the flag location indicates that the buffer is in use; a zero value indicates that the buffer is free):

CPU 1 wants the buffer and reads the flag location to see if the buffer is free.

CPU 1 finds that the flag is 00, indicating that buffer is free.

CPU2 wants the buffer and reads the flag location to see if the buffer is free.

CPU 2 finds that the flag is 00, indicating that the buffer is free.

CPU 1 writes 01 to the flag location, indicating that it is taking the buffer.

CPU 2 writes 02 to the flag location, indicating that it is taking the buffer.

Now we have a conflict—each CPU thinks it has control of the buffer. But now we will add the second arbitration step:

CPU 1 checks the flag location again, finds that 02 is there instead of 01, knows it has lost the arbitration, and waits.

CPU 2 checks the flag location again, finds that 02 is there, and knows it has buffer.

Of course, you have to handle the case where CPU 2 is a little slow and writes the 02 after CPU 1 has performed the second check. This might occur if CPU 2 has a slower clock or gets an interrupt between wanting the buffer and asserting control of the buffer. One way around this is to have a sufficient delay between writing and checking the buffer to ensure that all the writes are finished.

Another way around the contention issue is to have a three-value flag and interlocked handshake. Each CPU has a flag location for the common resource, and each is assigned a priority.

When one CPU wants the resource, it checks the flags for all the CPUs. Only if all the flags are 0 can it request the resource, by writing 01 to its own flag location. Then it checks all the flags *again*. If a higher-priority CPU has requested the resource (by writing 01 to its flag location), the lower priority CPU must wait. It indicates this by writing 02 to its flag location. If no higher-priority CPUs have requested the resource, it indicates ownership by writing 03 to its flag location.

If a higher-priority CPU wants the resource, it does the same checks before writing 01 to the buffer. If a lower priority CPU has written 01 at the same time, the higher-priority CPU cannot take the resource until the lower-priority CPU writes either 02 or 03. If the lower-priority CPU writes 03, the higher-priority CPU was a little behind and must wait. If the lower-priority CPU writes 02, then the higher priority CPU can write 03 and take the resource.

This complicated scheme is needed because there always is a possibility that one CPU will write 01 to its flag location after the other CPU has read the flags, found them 0, and written 01. The following are four possible contention scenarios and how this protocol handles them (CPU 2 has the highest priority in all these examples). In Scenario 1,

CPU 1 checks the flags and finds them all 0.

CPU 2 checks the flags and finds them all 0.

CPU 1 writes its flag to 01.

CPU 2 writes its flag to 01.

CPU 1 checks the flags again and finds that CPU 2 has set the flag.

CPU 2 checks the flags, finds that CPU 1 has set the flag, and waits to see what CPU 1 will do.

CPU 1 sets the flag to 02, indicating that it will wait.

CPU 2, polling the flags, sees CPU 1 indicate that it is waiting.

CPU 2 sets its flag to 03, indicating that it is taking the resource.

In Scenario 2,

CPU 1 checks the flags and finds them all 0.

CPU 2 checks the flags and finds them all 0.

CPU 1 writes its flag to 01.

CPU 1 reads the flags again and finds that its flag is the only one set.

CPU 2 writes its flag to 01.

CPU 1 writes 03 to its flag, indicating that it is taking the resource.

CPU 2 checks the flags again, finds 03 in CPU 2's flag, and waits.

In Scenario 3,

CPU 1 checks the flags and finds them all 0.

CPU 2 checks the flags and finds them all 0.

CPU 2 writes its flag to 01.

CPU 2 checks the flags and finds its flag is the only one set.

CPU 2 writes 03 to its flag, indicating that it is taking the resource.

CPU 1 writes its flag to 01 (CPU 1 was slow or delayed).

CPU 1 checks the flags, finds that CPU 2 has taken the resource, and waits.

In Scenario 4,

CPU 1 checks the flags and finds them all 0.

CPU 2 checks the flags and finds them all 0.

CPU 2 writes its flag to 01.

CPU 2 checks the flags and finds its flag is the only one set.

CPU 1 writes its flag to 01.

CPU 1 checks the flags and finds that both flags are set.

CPU 2 writes 03 to its flag, indicating that it is taking the resource.

CPU 1, having lower priority, writes 02 to its flag and waits.

Of course, when a CPU is done with the resource, it must always reset its flag to 0 so the other CPUs know the resource is free.

Engineering Specifications

This was mentioned briefly in Chapter 1. Although not a requirement for most designs, the engineering specifications is a document I have found useful for large, usually multiprocessor designs. This document can cover the entire system, including mechanical design, or just the electrical and software part of the design. The engineering specifications should include the following:

A brief description of the product.

A description of how the design will be accomplished. This includes what parts of the design will be new and what will be reused from old designs.

Functional breakdown of the software and hardware. This includes what boards will be used, which functions they will perform, what processor family will be used.

Interface definition. Interfaces to the outside world should be defined in the requirements document so they need to be only summarized in the engineering specifications. The interfaces between processors, both electrical and software, should be described in detail.

Board requirements for each board.

Software requirements for each processor, where appropriate.

The goal of the engineering specifications is that, from it, any engineer could implement the design. While this level of description rarely is achieved in practice, it is a good target to aim at. The table of contents for a generic engineering specification might look something like this:

Scope

Design approach

Existing components that can be reused

New designs required

Electrical system block diagrams

Subcontract work

Electrical architecture

Functional breakdown—board level
Interboard/interprocessor communication interfaces

Software architecture

Interprocessor communication interfaces
HLL to be used

Board requirements documents

Software requirements documents

Since we focus on embedded systems in this book, mechanical design is ignored in this example. However, any electromechanical system also would require an equivalent section for mechanical design.

Chapter 8 provides an overview of real-time operating systems.

Real-Time Operating Systems 8

The theory and use of a real-time operating system (RTOS) can and has taken entire books. This chapter provides an overview.

As embedded systems grow in complexity, they start to look more and more like their personal computer (PC) cousins. Software development for an embedded system often is complicated by the need to control system resources. In addition, some embedded systems need to connect to Ethernet interfaces, hard disk drives, and other PC-like peripherals. If all the software is written from scratch, then code must be written to interact with every device. For many standard interfaces, this is a duplication of the effort already expended by some other software engineer, somewhere.

In a typical embedded system, each function or process handles its own resources, somewhat independent of the others. A process that interfaces to a host system over an Ethernet link, for example, has memory allocated for its buffers. A similar process may have code and buffers for an RS-232 connection. The polling loop gives each process control, one at a time, and each checks for data to or from its respective interface. But say that, in this example system, the host uses either Ethernet or RS-232, never both. In that case, the system really does not need both pieces of code and both sets of buffers active at the same time. This system could get by with less RAM by managing the memory, allocating whichever buffer is not needed to other purposes.

In addition to memory management, all embedded systems must schedule processes in some manner. The polling loop method, sometimes called *sequential* or *round-robin scheduling*, is probably the most common. In the pool timer example, each task (motor on/off control, time rollover handling, keypad processing) is given control one at a time. When motor control is finished, it passes control to the time rollover process, which subsequently hands off to the keypad (mode control) code, which returns to the motor control code. Task scheduling is one big loop.

Although this method works well for simple systems like the pool timer, it has some drawbacks. In the pool timer example, each task runs until it is finished. The keypad processing code takes as much time as is needed to handle user inputs from the keypad. Again, these tasks are very simple and the longest processing time for the most complicated task is still too short to be a problem.

But imagine a system that is controlling an automated assembly line. There might be code that sorts the incoming material, adjusts the temperature of processes, regulates the speed of the motors that move objects down the line, and tests the finished products, rejecting any that are bad. In such a system, the temperature control might have a fairly long delay, so it could take a while to get the temperature right. If the temperature routine sets the temperature and then waits to see what happens, all the other functions are held up in the meantime. In other words, the processing time for one task affects the ability of others to do their jobs.

A second problem with sequential task ordering is that all tasks have the same priority. In the assembly line example just mentioned, imagine that the assembly line gets jammed. The code that handles the jam and shuts down the line should take priority over everything else.

Actual sequential scheduling systems, of course, do not really assign tasks that way. The temperature process would not keep control of the system but would adjust the temperature and check it again the next time it is executed. But the concept still is valid—handling a jam may take priority over the temperature, regardless of how far out of tolerance it is.

A third potential problem with sequential scheduling is the sheer number of tasks. If the number of tasks in the system is too large, it may be impossible for the system to keep up with processing demands, even if each individual task takes little time. Each task in a sequential arrangement requires a certain amount of time to execute, even if it is just checking to find out that it has nothing to do. The communication protocol converter mentioned in Chapter 3 is an example of this. The output code checks for buffer not empty. If the buffer is not empty, it proceeds to check for interface ready. If the interface is ready, it sends a byte. If it has nothing to transmit or if the interface is not ready, the code passes control to the next task. But even if the process cannot send because there are no data or because the output device is not ready, checking for these conditions takes time.

That protocol converter had four very simple tasks: receive data processing, XOFF processing, output processing, and XON processing. One way to handle task scheduling would be to have each task active only when needed. Receive data processing might get a byte and put it in the first in, first out (FIFO) buffer. It then activates the output task. XON and XOFF are inactive. So the program loop transfers control from receive to output and back, skipping over XON and XOFF and their minimal checks. Then suppose that enough receive data are placed in the buffer to require an XOFF be sent to the host. The receive process detects this condition and activates the XOFF process. The XOFF process remains active, waiting for the interface to the host system to be ready and then transmitting the XOFF byte. The program loop would then be receive-XOFF-output-receive and so on. Once XOFF completes its task (sending XOFF), it deactivates itself and the loop returns to the receive-output sequence. If the output code empties the FIFO buffer, sending the last byte to the output device, the output code deactivates itself until more data are available. Figure 8.1 illustrates this process.

Suppose that the system were more complicated and the return link to the host was used to send other data in addition to the XON/XOFF flow control. Since sending XOFF is a high priority (failing to do so risks buffer overflow and missed data), XOFF may be activated as a higher priority than any other serial output task. This makes sure that the XOFF code gets the next available transmit slot on the serial interface.

The protocol converter admittedly is much too simple to benefit from such a scheduling system. The code to handle scheduling would be longer than the code just to do a sequential loop. But in complex systems, using an RTOS provides just this type of scheduling capability.

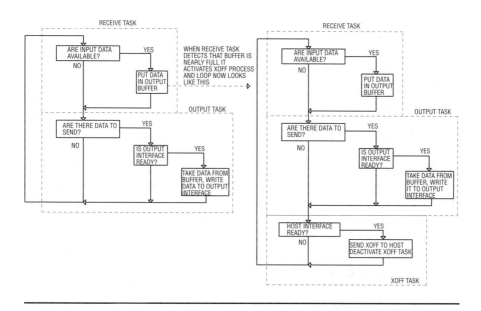

Figure 8.1
Communication System with Scheduling Implemented.

Like the operating system in your PC, an RTOS (sometimes called a *real-time executive* or *real-time kernel*) manages the limited resources of an embedded system. Your PC does not keep every program on the disk in memory at once. Programs are loaded and executed only when you select them.

Real-time operating systems have one characteristic that is key to use in real-time designs: They are deterministic. That is, the vendor supplies you with information as to how long it takes to perform specific operations, such as activating a task. Knowing this, you can predict the impact the RTOS will have on system performance.

An RTOS comes in two basic flavors: kernels and full operating systems. A kernel usually implements the basic task and memory management functions. A full operating system may have drivers for disk drives, serial ports, and other common resources. One common characteristic of RTOSs is that the system hardware must generate a regular interrupt (called a *timer tick* or just a *tick*), say, at 20 Hz (50 ms). This is used for timekeeping, task scheduling, and other functions. Not all RTOSs require a system timer interrupt. Real-time operating systems typically support the following functions:

Multitasking, which includes

   Activation and deactivation of tasks
   Setting task priorities
   Scheduling tasks

Communication between tasks

Memory Management

Multitasking

This is the process of scheduling tasks or processes so that they all appear to operate simultaneously. In the protocol converter example, the receive, XON, output, and XOFF processes appear to a human user to run simultaneously, because the sequential, one-at-a-time operation is so fast. All the functions of task activation/deactivation, scheduling, and ranking are part of the multitasking function. In a sequential program, none of these operations, which are required for true multitasking, is implemented.

Multitasking also can be implemented by *time slicing*. In this method, tasks are switched every tick. Every time the interrupt occurs, a different task is given control, so each task gets to execute for one tick time (50 ms for the 20 Hz tick example given previously). The overall execution speed for a

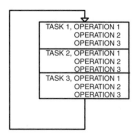

SEQUENTIAL SCHEDULING:

Say there are three tasks, each
with three operations to perform.
In sequential operation, each
task runs until finished.

TASK 1, OPERATION 1
OPERATION 2
OPERATION 3
TASK 2, OPERATION 1
OPERATION 2
OPERATION 3
TASK 3, OPERATION 1
OPERATION 2
OPERATION 3

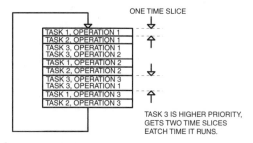

TIME SLICING:

The same three tasks. Each
is given specific time slices.
Each task runs only until its
time slice is up.

ONE TIME SLICE

TASK 1, OPERATION 1
TASK 2, OPERATION 1
TASK 3, OPERATION 1
TASK 3, OPERATION 2
TASK 1, OPERATION 2
TASK 2, OPERATION 2
TASK 3, OPERATION 3
TASK 3, OPERATION 1
TASK 1, OPERATION 3
TASK 2, OPERATION 3

TASK 3 IS HIGHER PRIORITY,
GETS TWO TIME SLICES
EATCH TIME IT RUNS.

Note: For simplicity, each operation is
one time slice in length. In an actual
system, the operations would be of
varying lengths and would be
halted in mid-operation at the end
of a time slice.

Figure 8.2
Sequential versus Time-Sliced Operation.

given task depends on the number of tasks. Higher-priority tasks can be allowed to execute for more than one tick time. A task that needs less than a full tick to execute can terminate early, giving the remainder of its time to the next task.

Sequential scheduling and time slicing are essentially the same, except that sequentially scheduled tasks run until finished and time-sliced tasks run until their time is up. Tasks operating under either scheduling scheme can voluntarily relinquish control before finishing. In that case, they can be restarted where they left off instead of starting over.

Most RTOSs can support time slicing or sequential scheduling. Sequential scheduling also can check for and stop tasks that hog the CPU. In any scheduling system, of course, only one task at a time actually has control of the CPU. Figure 8.2 illustrates the difference between time slicing and sequential operation.

Preemptive Scheduling

Preemptive scheduling is the most common method of scheduling tasks when using an RTOS and is one of the primary advantages of using an RTOS. Under preemptive scheduling, a task runs until it is finished or until a task of higher priority preempts it. Before going into more detail about preemptive scheduling, we should look at RTOS's task handling in general.

Activation and Deactivation of Tasks

Tasks under RTOS can be ready or not ready. The RTOS keeps a list of tasks that are ready and what their execution priority is. A ready task is added to the task list and executed in sequence. When a task becomes not ready, it is removed from the list. Going back to the protocol converter example, the output task might go ready when there is data in the FIFO buffer and become not ready when the FIFO buffer is empty.

A ready task may be inhibited from running because something blocks further execution. For example, the protocol converter output task may be ready because the FIFO buffer is not empty but blocked from doing anything because the output interface is not ready for the next byte. The RTOS-based system software in this case might deem the output task not ready (suspended) and replace it with a task that checks for output ready. When the output is ready, the check-for-output-ready task removes itself and deems the output task ready. Of course, this makes sense only if the output checker task takes less time to execute than the normal output task takes to check for output ready or if suspending the output task frees up the output interface for another device to use. The output device might be ready for one kind of data but not ready for another, so it can be used by another process instead of sitting idle.

Once given control, a task may run until it is finished or until it finds that it cannot execute further, like the output task condition just mentioned. In either case, the task transfers control back to the RTOS, which then passes control to the next task in sequence. In this respect, the RTOS-based system is like any sequential execution system, with the added ability to remove tasks from the sequence of execution. When a task is activated, the priority may be set at the same time, depending on the specific RTOS used.

Event-Driven Scheduling

The practice of adding and removing tasks from the task list based on changing circumstances is called *event-driven scheduling* and, with preemptive scheduling, is the method used in many, if not most, RTOS-based systems. Preemptive scheduling more closely models the real world; you might plan to go to work today, but a fender bender on the way will change your priorities, at least until the police report is finished. Once you get to work, you might have scheduled a project meeting, but an emergency staff meeting called by your boss takes priority.

In a preemptive, event-driven system, an event such as an interrupt or a task may determine that some other task needs to be activated. It may do this, for example, by setting a semaphore or placing data in a mailbox. The task,

Embedded Microprocessor Systems

which was previously set up to be activated by the RTOS when this event occurred, is activated if it has a higher priority than the current task.

If the protocol converter were preemptively scheduled, the priority might look like this:

Receive processing (highest priority)

XOFF processing

XON processing

Output processing (lowest priority)

A single interrupt, generated by a byte in the serial input register, might drive the system. The receive task might ask to be activated when a particular semaphore is set. When a byte is received, the receive ISR (interrupt service routine) sets this semaphore and the RTOS activates the receive task. The receive task reads the byte from the universal asynchronous receiver/transmitter (UART) register, processes the data, and places it in a buffer for the output task. The receive task has the highest priority because the system cannot afford to miss a serial input byte. Once the receive task has finished processing the received byte, it becomes not ready until the next receive interrupt occurs, by asking the RTOS to suspend it until the semaphore is set again.

When the buffer is near full, the XOFF task is activated (becomes ready) by the receive processing. XOFF could be "created" by receive processing, where the receive processing requests that XOFF processing be activated, or XOFF could have previously been set up to be activated by a semaphore like receive processing was. XOFF runs until it has successfully sent the XOFF signal to the host or until it is preempted by the receive task. If receive preempts XOFF, it gets control (from the RTOS), processes the receive data, and then control is returned to XOFF. Again, the highest priority ready task is the one executed.

XON is next in priority. If output processing empties the buffer past a certain point, it activates XON. Output processing has the lowest priority, which is possible because the XOFF task prevents the buffer from overflowing, so no data ever are missed. Of course, if the receive data flow cannot be suspended with XOFF, then output processing would have to have a higher priority so the buffer did not overflow.

None of this happens by magic. The RTOS can activate a task, when a semaphore is set or a message is received, only if it was previously told to do so. Also, in a RTOS-based system, the interrupt service routines usually get control via the RTOS, so an ISR may not need to set a semaphore to start a task. Instead, the RTOS can schedule the task upon activation of the ISR itself.

A final note about scheduling: Both sequential and preemptive scheduling systems allow a task to run until finished. The difference is that, in a preemptive system, a task runs until finished or until preempted. Between two ready tasks of different priorities, the higher-priority task always preempts the lower-priority task and finishes first. If two tasks of equal priority are ready at the same time, a sophisticated RTOS usually activates the one that has been idle the longest.

A note about terminology: A task is considered active when it actually is running, when it has been given control of the CPU by the RTOS. A ready task is in the list of tasks waiting to run. A task can be not ready, such as the receive processing on the protocol converter while waiting for the start semaphore.

The RTOS knows about only those tasks it has been told about (been created). The code for other tasks may reside in memory, but they are invisible to the RTOS until they are created.

Keeping Track of Tasks

The RTOS keeps track of tasks with a *task control block* (TCB). This is where the RTOS saves information about tasks. One TCB entry is made for every tack managed by the RTOS. The TCB must store the following:

- **Task ID.** Typically, the task number. Depending on the RTOS and the processor it is running on, there may be a maximum of 128 tasks, 256 tasks, 32,768 tasks, and so forth. The maximum number of tasks usually is what can be identified with a byte/word/doubleword or whatever the word width of the processor registers.
- **Task state.** Ready, blocked, and the like.
- **Task priority.** The priority level of the task; a numerical value, usually 0–127, 0–32,767, and so on.
- **Task address.** Where in memory the code for the task is located.
- **Task stack pointer.** The microprocessor stack is used to pass variables and store the context for subroutine calls and interrupts. Each task needs to be able to perform subroutine calls and service interrupts (or at least save the return address for an interrupt). For this purpose, each task has its own stack. The TCB includes the value of the stack pointer when the task last executed (or the top of the task stack the first time the task executes).

The task stack is the *microprocessor stack*. As each task is given control, the microprocessor stack pointer is modified to point to the stack for that task. Each task must be allocated sufficient stack space to save the processor context, any dynamic/temporary variables stored on the stack, and any information stored to the maximum depth of subroutine calls. The processor context also may need to include things such as the context (registers) of a floating-point coprocessor or something similar. When a task stops running for any reason, the RTOS stores the stack pointer for the task in the TCB.

Once a task is ready to run again, the processor context needed to resume execution where it left off is stored on the task stack. The RTOS just has to get the stack pointer from the TCB, put that value into the microprocessor stack pointer, and return control to the task.

Depending on the RTOS, the TCB may contain additional information such as environment information for dynamically allocated tasks and the like.

In addition to a stack for each task, the RTOS will have a kernel stack for use by the RTOS itself.

Communication Between Tasks

In the protocol converter example, the receive task puts output data in a common FIFO buffer. If an RTOS is used, the data normally are passed through the RTOS. The RTOS may support semaphores, buffers, queues, and mailboxes.

An RTOS semaphore is similar to the key-press flags or semaphores used in the pool timer. A task asks to set the semaphore and another task can wait on the semaphore or possibly reset it. Setting a semaphore can activate a task. The difference in an RTOS system is that all access to the semaphore is passed through the RTOS.

An RTOS buffer is just like the FIFO buffer used in the protocol converter, except that the RTOS manages it. If the protocol converter uses an RTOS, the receive task requests a buffer from the RTOS, puts the data to be transmitted in the buffer, and tells the RTOS to pass the data to the output task. The output task typically receives pointers to the buffer, telling it where the buffer is in memory and how many bytes (or words or whatever) are there.

A queue is a string of buffers. If the protocol converter worked on a message basis, outputting data only when a complete message is received, a queue could be used for this. The receive process could place a message in a buffer and pass the buffer to the output task. The output device might be busy when the next message is ready, so the receive task asks the RTOS to put the next message into a queue, where the output task processes the messages in the order they were received once the output device becomes ready.

In a mailbox structure, a task typically receives mail from several other tasks, just like you do at home. The RTOS manages the mailboxes, storing messages for a task until the task is ready to read them. Like a physical mailbox, once a task sends a message, it cannot take it back. Depending on the RTOS, a task may check for mail and wait if there is none, like you did when you were a kid expecting a package to arrive. An RTOS usually supports multiple mailboxes per task, as if you had a mailbox at home and several boxes at the post office.

Figure 8.3 summarizes RTOS communication. In the first diagram (RTOS buffer), Process A transfers data to Process B via a buffer. In the second diagram (RTOS queue), Process A filled buffers (queues) 1 and 2 and is filling buffer 3. Process B is taking data from buffer 1.

The third example in Figure 8.3 is of an RTOS mailbox. Processes A, B, and C are placing data in a common mailbox for Process D. Each message from each process is stored separately, like physical letters, each in its own envelope. Like when you sort through your mail at home, opening important

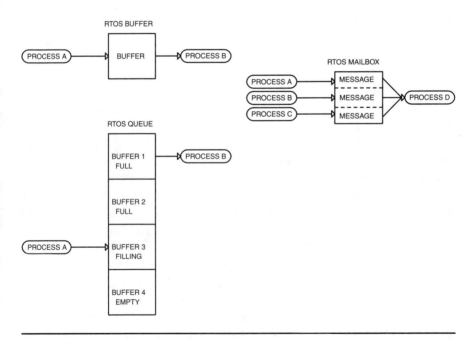

Figure 8.3
RTOS Communication.

letters first, the RTOS typically allows the sending process to assign a priority to the message for the receiving task.

Scheduling Tasks

A task may be ready immediately after an event occurs or it may be scheduled to start later. As already mentioned, a task may be scheduled to start when a semaphore is set, perhaps as a result of a hardware event. It also may be scheduled to start after a number of ticks have elapsed or at a specific time of day in systems that maintain time of day.

Memory Management

As mentioned in the section about buffers and queues, a process requests memory from the RTOS when it needs a buffer. This allows the system to get by on less memory than otherwise would be required.

In the first example mentioned in this chapter, a system used either Ethernet or RS-232 to communicate with a host PC. Say that receive data needs 256 K of memory for each interface and transmit needs the same. In a non-RTOS system, the Ethernet and RS-232 codes might each allocate 512 K (256 K receive, 256 K transmit) for a total memory requirement of 1 MB. But, as already mentioned, only one interface is used at a time. In a RTOS-based system, if each task requests only those buffers it actually needs, only one task will ever request buffers, making the memory requirement 512 K. Furthermore, suppose the system is half-duplex, meaning that transmit and receive never occur simultaneously. In that case, the receive task allocates a buffer, the data is processed, and then the transmit task is activated and it allocates a buffer. Since both buffers are never active at the same time, only 256 K of memory is needed.

The RTOS typically allocates memory as blocks, chunks of contiguous memory of a minimum size. If the block size is 1 K, for example, a task that needs a 14-byte buffer has to request a block and will get 1 K allocated to it. Determining block size is important in the system design. A task that needs multiple blocks usually needs the memory to be contiguous, so the RTOS has to find sufficient contiguous blocks of memory to meet the request. If blocks are too small, memory can become fragmented, because blocks are not necessarily released by the task in contiguous order. On the other hand, if blocks are made too large, there will be too few blocks to meet all the memory requests of all the tasks. Figure 8.4 illustrates both problems in graphical form.

In Figure 8.4, a small memory is shown. In the first half of the figure, when memory blocks are too small, memory becomes fragmented. Task 1 has allocated three blocks and then gave two back. Task 2 did the same. Task 3 has four blocks allocated, Task 4 has two blocks. Now if a fifth task needs three contiguous blocks, there is a problem. Blocks 1, 2, 5, 7, 8, 13, and 16

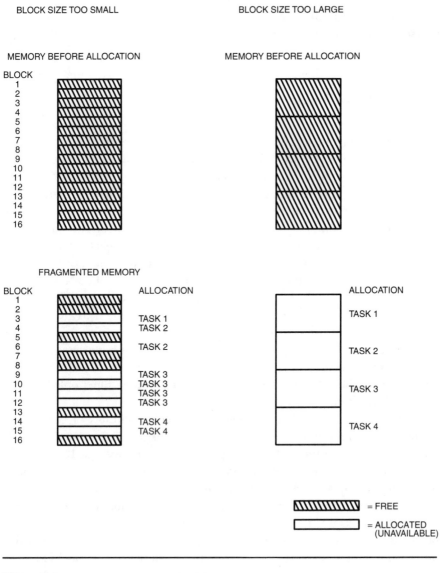

Figure 8.4
RTOS Memory Allocation Blocks.

are free, but no three of these are contiguous. Task 5 cannot get enough memory to run.

The second half of Figure 8.4 shows the opposite problem, when blocks are too large. Tasks 1 through 4 have each allocated one block, even though each task may need only a small portion of the allocated memory. When a fifth task needs a block of memory, none is left.

Resource Management

Say our protocol converter has two possible types of data for the output device. Perhaps, in addition to the normal receive-to-output path, diagnostic messages also are sent to the output. In this case, an RTOS might manage the output interface resource. If the normal output task needs to send received data, it requests access to the output interface from the RTOS. The RTOS grants the request, and the output task starts sending data. In the meantime, the diagnostic task requests the interface as well. Since the interface already is allocated to the output task, the diagnostic task has to wait until the interface is released by the output code.

In the area of timers, the RTOS may provide system timers, implemented in software and counting timer ticks, that can be allocated like any other resource. These timers, of course, cannot time anything with a resolution less than the timer tick interval.

RTOS and Interrupts

Obviously, an RTOS needs to handle interrupts if it is to manage a real-time operating system. When an interrupt occurs, the processor hardware handles the interrupt as it normally would, saving the return address and vectoring to the ISR. Assuming that a task is running when the interrupt occurs, the return address is saved on that task's stack. An RTOS usually will have special kernel services for interrupts. When the ISR gets control, it can use these special services. The simplest ISR servicing technique is to set a semaphore, do whatever must be done to reset the hardware that generated the interrupt, and then exit. Consequently, an RTOS usually will provide at least three kinds of services for ISRs. The first service, interrupt entry, allows the ISR to notify the RTOS that the interrupt occurred. The interrupt entry routine may save the processor context or other information and may be a generic routine provided by the RTOS vendor. The second ISR service is to request a semaphore set. The third service is an exit service call that notifies the RTOS when interrupt servicing is complete.

Special RTOS services are provided for ISRs because the ISRs cannot use the normal RTOS services. The normal RTOS services usually do not allow reentry. If an interrupt occurs while the RTOS is executing and the ISR attempts to use the RTOS function that was executing, disaster usually will result.

When the ISR exits (via the RTOS), the RTOS may perform a task switch, giving another task higher priority than the one that was interrupted. In the serial communication example, the receive interrupt may cause the RTOS to switch to the receive task. When the receive task is done, the RTOS can return control to the interrupted task. Since the processor context was saved on the task stack, resuming the task after an interrupt essentially is the same process as resuming a task that has become unblocked.

Typical RTOS Communication

Every RTOS is different, but the following is a list of RTOS services that would be typical (I made up descriptive names):

Define_Task. This defines a task to be executed. The typical parameters passed to the RTOS might include the task number, priority, and the task entry address.

Activate_Task. Requests activation of a task. The parameters passed to RTOS would include the task number.

Deactivate_Task. Deactivates a task. The parameters would include the task number.

Yield. Tells the RTOS that the task is finished for now and that the next task on the list may be executed.

Define_Timeslice. Defines the number of time-slice intervals that the task will be allowed to execute.

Allocate_Memory. Requests a specified number of memory blocks.

Mailbox_In. Receive a mailbox message. The parameters would include the task number and the mailbox number.

Send_Mail. Send mail to a mailbox. The parameters could include the mailbox number, destination task number, and priority of the message.

Wait_On. Waits for the queue to fill, the semaphore to be active, or the mailbox to receive mail.

Of course, this is not a comprehensive list of RTOS services, just an indication of the kind of things a RTOS supports.

A few pointers if you are thinking about using a RTOS: Make sure that assumptions about memory are correct. The Ethernet/RS-232 system assumed that transmit and receive were half-duplex. If this assumption turns out to be wrong; if both buffers are ever needed simultaneously, then there will be a memory allocation problem. This may be a minor problem, as a task waits until the memory is available. But, it can cause a lockup if the task that has the memory will not release it until the task requesting memory can execute.

Make sure the RTOS does not bog down the system operation. While an RTOS is deterministic, it still takes time to do things. Be sure this time is no problem.

Make sure that task priorities do not prevent a low-priority but essential task from ever executing.

Preemption Considerations

Two considerations you have to keep in mind when using a preemptive RTOS is that the RTOS manages the operation of the software and any RTOS function can perform a task switch. The idea is to get maximum use of the CPU, but it means you have to take things into consideration that otherwise you need not. Say you have an ADC that requires you to read the result within 100 microseconds (μs) of starting a conversion. Also say you have a solenoid that is activated by the software, held for 20 ms, and then turned off. The solenoid timing is performed by counting interrupts from a 1 ms timer. The polling loop activates the solenoid, and sets a variable, SOLENOID, to 20. The 1 ms ISR decrements SOLENOID as long as it is nonzero. When it decrements to 0, the solenoid is turned off. After the solenoid is turned off, a pump is started.

The way this might work in a polled environment is:

```
Activate solenoid
Start solenoid timer by setting SOLENOID to 20
Poll for 19 ms, checking to see if SOLENOID went to 0
Polling loop finds that it is time to start an ADC conversion.
Call ADC routine
    Start ADC conversion
    Wait for ADC to complete
        *** Right here, the timer interrupt occurs, so the ISR
        decrements
```

SOLENOID. SOLENOID decrements to 0.

Read/store ADC result

Return to polling loop

Polling loop checks SOLENOID, finds it has rolled to 0, turns off solenoid.

Now, in a RTOS environment, it might work like this:

Solenoid/pump driver task turns on solenoid and suspends for 20 ms.

19 ms go by, during which other tasks are executed

Some event tells RTOS that it is time to start an ADC conversion

RTOS starts ADC conversion task

ADC conversion is started

*** Again, the timer interrupt occurs, and the RTOS finds that 20 ms has gone by.

RTOS reactivates solenoid/pump driver task

Solenoid is turned off

Pump is turned on

Other processing goes on until solenoid/pump drive task suspends again.

RTOS reactivates ADC task, but now it is too late. ADC result is bad.

The result here is that, sometimes, the ADC conversion will be bad. There are a number of ways to fix this. The ADC task could be given higher priority than the solenoid/pump driver task. Or, before starting the conversion, the ADC task could tell the RTOS that it is about to begin a noninterruptible function (if the RTOS supports that). Or, the ADC task could ask temporarily to have its priority set higher than the solenoid task until the conversion is complete (again, assuming the RTOS supports it). The point is that, in an RTOS environment, any event that results in an RTOS function being executed can result in a task switch. An ISR does not necessarily return to the task that was executing when the interrupt occurred. You have to take this into account in the software. You do not know when interrupts will occur, so you have to assume they will occur at the worst possible time.

Use of an RTOS also can affect the hardware. In the previous example, the ADC conversion was assumed to be polled in some manner by the ADC conversion routine. In a preemptive RTOS environment, it may make more sense to have the ADC cause an interrupt when conversion is complete, allowing the ADC read to operate at the (high-priority) ISR level.

Applicability of RTOS

An RTOS is not suited to every application. Specifically, an RTOS probably is not a good solution if the device has to execute very high-speed interrupts, such as a low-level motor controller, or if the system is simple enough to make an RTOS more work than a simple sequential or state machine design. This does not preclude the use of an RTOS if an occasional interrupt occurs that requires immediate service, but the closer the processor is to bit-level control of the hardware, the less sense a RTOS usually makes.

An RTOS typically *is* used where the system needs shared resources, needs to allocate memory, or where operation is at a sufficiently high level to justify the RTOS overhead. In general, if the system is complicated but tasks can be scheduled at the resolution of the timer tick, an RTOS may make sense. Even in simple systems, an RTOS may be used to structure code execution. An RTOS also makes sense if you need standard resources (disk drives, VGA display, etc.) for which you want off-the-shelf drivers.

An RTOS is suitable any time the number of tasks is such that sequential scheduling is unable to ensure that the highest-priority jobs get done first. Using preemptive scheduling, an RTOS can make sure that the important functions get done on time.

Most RTOSs are configurable—you start with the basic kernel and add whatever features you need. If you have disk drives, you might add the RTOS module that includes disk drivers. Ethernet support or TCP/IP might be another module.

When you consider an RTOS, look at the cost. Some RTOSs have a one-time purchase fee, while others charge a license for every copy used. Sometimes you pay a sliding license fee, starting with a basic fee for the kernel and increasing as you add RTOS modules (such as TCP/IP support). License fees can get quite expensive, especially if your system has multiple processors needing an RTOS.

While the division between an RTOS and a kernel is not a fine line, generally a kernel is smaller than a corresponding RTOS. While not providing all the features of the full RTOS, the kernel can provide scheduling and management functions suitable for small, embedded systems that cannot support or do not need the overhead of a full RTOS.

Using an RTOS often means more memory, since each task will have its own stack. Some RTOSs are linked into your code, while others are like a PC OS: The RTOS loads from a storage device and your program runs as an application. Which RTOS you choose can have a big impact on the hardware; you need whatever basic resources the RTOS requires to operate.

Communication standards also are important. Many RTOSs now support TCP/IP, for example. If you use a standard interface such as this, you can communicate with any other system that uses the same standard protocol, regardless of what OS it uses.

Using an RTOS on a microcontroller presents special challenges: The code space is limited, the stack pointer may be implemented in hardware (making task stacks impossible to implement), and the RAM is very limited. Microcontroller RTOSs usually are very basic kernels, without many exotic features.

Chapter 9 describes some industry-standard platforms that you can use when designing an embedded system.

Industry-Standard Embedded Platforms

9

As mentioned earlier, one characteristic of an embedded system is that it is self-contained, requiring no user input to get started. There are some exceptions to that rule, which we look at here.

A problem with developing all parts of an embedded system is that all the interfaces—Ethernet, FDDI, RS-232—have to be developed along with the system. You have to design an interface circuit (or board) and cannot take advantage of off-the-shelf boards and driver software. One platform, however, allows you to use existing parts—the PC platform, in this case, the IBM PC/AT and its derivatives.

If you design an embedded system around a PC, you can get interface boards, disk drive interfaces, A/D and D/A interfaces, and a number of other components from existing vendors and often with driver software.

Advantages of Using a PC Platform

There are a number of reasons why some developers choose the PC platform for development.

Speed of Development

An embedded system designed from scratch requires that boards be designed, fabricated, and debugged. The software has to be tested and debugged on the target system. If a PC platform is used, the boards are available and the software can be written and debugged in the same environment. In addition, PC hardware with high-speed buses like peripheral component interconnect (PCI), take more design effort to get right. If you buy the boards, someone else has done the job of making them work.

215

Development Cost

Embedded systems based on a PC platform require no costly board design/fabrication/debug cycles. PC tools usually are used for software development, eliminating the need to purchase emulators. As product development cycles get shorter, there is an incentive to buy proven, off-the-shelf components. Another factor driving the use of purchased hardware is increasing clock speeds. As CPU speeds approach half a gigahertz, it is increasingly difficult for every company that needs a processor board to create its own designs. The tools are prohibitively expensive, partly since extensive simulation is required to ensure a good design.

Specialization

Some embedded designs still can be accomplished using processors with clock rates in the low megahertz range. But, as clock rates go up and development costs follow, more companies concentrate their efforts on the hardware and software that makes their products unique. Off-the-shelf CPUs, Ethernet boards, and similar components are treated as commodity parts, which they are. This is buying the "jellybean" parts of the design, leaving the company's engineers free to do the unique things. Since all modern, high-speed CPU boards essentially are the same, you pick a CPU, pick a chipset that supports it, and wire it accordingly. Why assign an engineer to spend three months developing a board that looks and works like a hundred other nearly identical designs?

Mass Storage

Disk drives, interface boards, and driver software are standard parts of the PC platform. Some systems need mass storage to capture data; for example, a system that keeps a log of instrument readings from a fluid pipeline. If the system takes a reading every second, the storage requirements can add up quickly. Other applications where mass storage could be a requirement include storing bitmapped images and store-and-forward interface systems. And some real-time operating systems are designed to operate with mass storage.

Standard Software

You need not learn the interface to a RTOS with a PC-based system, as DOS (or OS/2 or Linux) is already available. Off-the-shelf software is available for communications, graphics display, and many other applications. New features, depending on what they are, may be bought instead of designed. If your application needs some kind of database, you can buy a database package instead of writing one.

Standard Hardware

Off-the-shelf interface boards simplify hardware design. Boards that need software drivers usually come with them, simplifying development.

User Interface

If the application needs a graphics display or keyboard input from the user, a PC already has the pieces in place.

Development Environment

Standard debugging software is available. The development language is not limited by the hardware. A huge base of development software is available from a number of vendors. On the hardware side, the PC ISA (industry-standard architecture) bus is well defined and easy to interface to. Even the PCI bus is a known standard, although more difficult to design for.

Flexibility

Adding features or options can be as easy as plugging in a board and adding the necessary software.

Easy Updates

Software updates involve loading new software from a floppy disk or CD-ROM. If a passive-backplane system is used, processor upgrades are simply a matter of plugging in a new CPU board (and, maybe, appropriate software).

Product Cost

If your product is manufactured in relatively low volumes, it can be expensive to build your own CPU boards and other system components. By using off-the-shelf parts, you take advantage of the volume advantage that the board vendor has. The vendor is selling the same board to numerous other users, so the total production volume can be high enough to make the purchased boards cheaper than boards built in-house.

CPU Hardware

With the PC architecture, you typically get a Pentium-class or better CPU. This brings with it all the Pentium-level hardware advantages, such as protected mode programming, hardware memory management, debug registers, and I/O protection. These improvements simplify multitasking.

Protected Mode Programming Intel ×86 processors, from the 386 on, implement protected mode programming. In the 8086/8088 and the 80186/188, the only memory model available is the Real mode. The addressing scheme for these CPUs permits addressing of 1 megabyte of memory (20-bit address), in 16 64-K segments. The 386 and higher CPUs use a different addressing scheme that permits access to 4 gigabytes of memory using 32-bit addresses.

Hardware Memory Management In the Real mode, every task has access to the entire 1-MB memory space. Any task can read or write to any address. In the Protected mode, the memory partitions may be protected so that a task cannot access memory outside its own segment. A segment of memory even can be shared between two tasks, so that one task can both write and read to the common area, while another task can only read it. Attempts by a task to execute or access memory outside its bounds cause the hardware to generate an exception condition that can be handled by the operating system.

Hardware Debug Registers The 386 and higher processors have registers that simplify debugging. The four address registers can be set to break on read, write, or read/write access. Debuggers can take advantage of these on-chip resources to simplify debugging.

I/O Control The ×86 architecture has a 64K I/O space, separate from the memory space, that can be accessed with unique instructions. In the Protected mode, the I/O space can be protected so that only the operating system has access to I/O ports. This forces applications to use the operating system resources to access I/O.

Drawbacks of Using a PC Platform

So, if it is this easy to design a PC-based system, why doesn't every system use a PC? There are a few drawbacks to using a PC.

Product Cost

This may not be an issue for low-volume applications or for systems where the embedded control components are a small part of a much larger system (such as an automated assembly line), but for consumer and other cost-sensitive applications it is. Imagine a microwave oven that has to be sold with a PC attached—enough said.

Another reason that drives product cost higher in PC systems is that you have to pay for everything that comes with the PC architecture. If your application needs no disk drive or keyboard, you still have those interfaces on the CPU board that you buy. CPU board vendors do not carry a large number of CPU boards, to fit every need. Since silicon is relatively cheap, they carry just a few boards that contain nearly everything a user might want. There is little choice about this, since these boards must be designed using off-the-shelf chipsets (discrete logic would be slow, expensive, and take enormous real estate). Most of the chipsets that interface to x86 family processors contain standard PC peripheral functions. The entire idea is to shrink the standard functions to the smallest size/cost possible for PC motherboards.

Hardware Development

For standard interfaces, off-the-shelf boards are available. But if a proprietary interface is required or if some unavailable function is needed, hardware has to be designed anyway. A distributed system, with low-level motor controllers and interfaces, probably has a PC as a central controller, and everything else is custom-made. The more hardware that has to be designed anyway, the less leverage an off-the-shelf CPU provides.

Keyboard and Monitor

The standard PC has a keyboard and monitor attached. They are bulky and sometimes unnecessary for the specific embedded application, but they have to be there.

Parts Availability

Try to buy a PC motherboard, and then try to buy the same motherboard a year later. This is nearly impossible to do—the designs just change too often. This can be a real problem, especially if every iteration of the design requires new EMI or safety agency investigation.

Not Real Time

PC operating systems, such as DOS, do not operate in real time. Some PC operating systems are multitasking, but that still does not mean they are real time. PC operating systems are not real time because they are not deterministic—you do not know how long an operating system function takes to execute. Some applications do not care if the operating system goes away for a quarter of a second to get something from disk; others do.

Mass Storage

This is an advantage if you need it. If you do not, you still need the disk drive from which to load your OS and your programs.

Design Problems

Buying an off-the-shelf CPU means someone else has verified the design. But, if subtle timing problems turn up in the hardware, you are dependent on the board vendor to admit they exist and fix them. You have no schematics, PLD equations, or the other information necessary to debug the design yourself. And you do not want to; that is why you chose to buy instead of build.

Some Solutions to These Problems

Some of these problems have been addressed and have solutions, but they all make the resulting system a little less compatible with the PC:

BIOS. Kits are available that allow you to write a basic input/output system (BIOS) that eliminates the keyboard, monitor, and other standard peripherals.

DOS in ROM. Development kits are available that allow DOS and your applications to be placed in PROM or flash memory, eliminating the requirement for a disk drive. But not all operating systems can run from ROM or without a disk drive. Annasoft makes products that will allow you to put DOS in ROM.

Passive backplane. The problem of parts availability sometimes can be solved by using a passive backplane. Essentially, this consists of the expansion slots from a PC motherboard without the motherboard. A CPU board plugs into one of the expansion slots. Other standard boards can plug into the other slots. While these backplane/CPU board combinations typically are more expensive than a clone motherboard produced to the tune of 100,000 per month, they solve the problem of not being able to buy the same board twice. But these boards are not perfect—they still depend on availability of parts, such as PC chip sets, that may go out of production.

RTOS. Real-time (that is, deterministic) operating systems that emulate DOS are available. Of course, all of them do not work *exactly* like DOS, which can cause problems. Some, however, are close enough to DOS that they advertise as being able to run Windows (or they did, before Windows 95/98 replaced Windows 3.1). One problem with using a non-DOS, non-Windows operating system is that you will not always find drivers for every peripheral chipset for every RTOS. For instance, you may be find that one vendor's motherboard uses an Ethernet chipset for which your RTOS vendor has no driver. Using an RTOS in a PC environment means you have to make sure there is a match between the PC hardware and the RTOS. And, if your hardware becomes obsolete, you have to be sure the new hardware is compatible, too.

Other options for embedded operating systems include Windows CE and the real-time version of Windows NT.

ISA-Based Embedded Boards

A quick perusal through a couple of catalogs from vendors who make ISA-based embedded boards includes the following:

- 486 CPU board with memory, integrated device electronics (IDE) disk controller, Ethernet interface, video graphics array (VGA) or liquid crystal display (LCD) controller, serial ports, printer port
- 48-channel digital I/O card
- 24-channel digital I/O card
- Digital I/O board with optically isolated inputs and relay closure outputs
- 16-input analog-to-digital (A/D) converter card
- 8-output digital-to-analog (D/A) converter card
- Watchdog timer card
- Synchronous serial port card

This chapter so far has focused on the PC as a platform for embedded systems. In addition to the problems already mentioned, a number of other problems with using a PC for embedded applications exist. First, the ISA bus is slowly going away, to be replaced by PCI, USB, and possibly Firewire or Bluetooth. PCI is much faster than ISA but more difficult to design for. As PCs need ever-faster peripherals, this transition makes sense. But many embedded applications, even those that require a very fast CPU, do not need high I/O speeds.

A PC is large and may be difficult to mount inside your product. Even the form factor of a PC motherboard is fairly large.

The average PC user will be running Windows 98, instead of a RTOS, and does not need to know how to write drivers for the chipset and peripherals on the motherboard. The embedded developer, on the other hand, needs this information; not being able to get it can make development difficult. Some PCs have a Plug and Play (PnP) BIOS that makes it hard to control how the interrupts and other features will be allocated.

Implicit in all these characteristics of the PC architecture is complexity. If you are building a PC-based product, you are virtually forced into using the BIOS on the motherboard and some kind of operating system. This is because the chipsets and peripheral functions on the board are complex enough (and sometimes proprietary enough) that it is impractical to write drivers and initialization code for them—unless you have an enormous development budget and a huge software team.

Finally, PCs are not intended for embedded application, so the only flash memory they have is for the BIOS; and you may not be able to find out how to program that. To load your code, you are stuck with having a hard disk or floppy drive that you otherwise might not need.

Other Platforms for Embedded Systems

PC 104 Bus

The PC/104 bus compresses the PC architecture to a form factor better suited to small embedded systems. The PC/104 bus is (almost) electrically identical to the ISA bus but on a different form factor. The PC/104, instead of using a backplane to interconnect the boards, has a *stackthrough* connector on each board. The pins on the back of one board connect to the socket on the front of the next. Two or more boards are stacked into a "sandwich" (see Figure 9.1). PC/104 boards are approximately 3.5 × 3.75 inches.

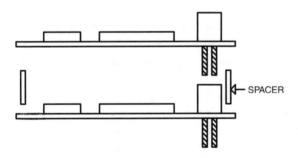

Figure 9.1
PC/104 Board Stacking.

The PC/104 bus comes in three versions. First is an 8-bit bus that closely matches the signals and timing of the original 8-bit IBM PC expansion connectors; the second is a 16-bit version that follows the 16-bit ISA connectors. The PC/104 signals have slightly different drive specifications, which correspond to their use in embedded systems, typically with a limited number of boards. A third version of the PC/104 bus is a PCI-like bus for high-speed transfers, the PC/104-Plus.

The primary drawback to the PC/104 form factor is also one of its biggest advantages—small size. Little room is left for connectors, and the board spacing prevents the use of large heatsinks for power devices. PC/104 CPU boards are available with processors ranging from an 80188 to 586- and Pentium-class processors.

One way that PC/104 CPUs can be used is as a smaller daughterboard on a larger I/O board. To drive a lot of motors, you might have a large board, filled with power ICs and motor drivers and controlled by a PC/104 CPU plugged into a connector in one corner.

Most PC/104 CPU boards provide a significant amount of flash memory, which usually can be configured as a virtual disk drive. This permits you to load an application and whatever operating system you use into silicon, with no need for a hard drive or floppy to get everything going. Many PC/104 CPU boards include an Ethernet connection, and you often can load the software directly from that. If your embedded controller is talking to an external PC via Ethernet, you can store the code in the PC and download it on power-up. This makes it easy to send software changes to the field.

Many manufacturers, such as Ampro, make CPU boards that are larger than the PC/104 form factor but retain the PC/104 interface connectors. This approach allows more room for components without giving up PC/104 electrical compatibility.

A drawback to using a PC/104 CPU is the same as using a PC: You may pay for stuff you do not use. This occurs for the same reasons it does on a PC—standard chipsets. Even if your application does not need VGA display, keyboard, or IDE interface, you probably get them on the PC/104 CPU anyway. You *might* be able to design a board without those features for less. But remember that the PC/104 manufacturer spreads development and production costs over a larger volume than you can. Some PC/104 manufacturers sell a depopulated version of their boards. If you are not using a VGA controller, for instance, they can leave off the video memory, making the board less expensive.

The introduction of the USB bus may help alleviate some of the size constraints on PC/104-based systems. Current PC/104 CPU boards typically are covered with connectors. Implementation of floppy, keyboard, printer, serial, and other interfaces takes real estate on the board. Even though these functions are embedded in complex chipsets, IC real estate still is used and inter-connections must be made. Connector space is tight enough that some PC/104 CPU boards require a floppy drive from a notebook computer (expensive), because no room is left on the board for the larger, standard floppy connector.

Although I have yet to see one produced, I can imagine a PC/104 CPU board that does away with the floppy, keyboard, printer, IDE, and maybe serial connectors, using USB instead. Such a board would be targeted at applications that do not need those peripherals except during development. During development, a "black box" could be used to interface the USB to all these standard peripherals. This black box could even be fairly expensive, since it would not affect product cost. During production, instead of having four to six unused connectors on the board, only the USB is unused. The board space preserved by this approach could be used for other interfaces or additional CPU functionality.

The pinout for the PC/104 bus is as follows:

PIN	J1/P1 (ROW A)	J1/P1 (ROW B)	J2/P2 (ROW C)	J2/P2 (ROW D)
0	—	—	GND	GND
1	−IOCHCHK	GND	−SBHE	−MEMCS16
2	SD7	RESET	LA23	−IOCS16
3	SD6	+5V	LA22	IRQ10
4	SD5	IRQ9	LA21	IRQ11
5	SD4	−5V	LA20	IRQ12
6	SD3	DRQ2	LA19	IRQ15
7	SD2	−12V	LA18	IRQ14
8	SD1	−ENDXFR	LA17	−DACK0
9	SD0	+12V	−MEMR	DRQ0
10	IOCHRDY	KEY	−MEMW	−DACK5
11	AEN	−SMEMW	SD8	DRQ5
12	SA19	−SMEMR	SD9	−DACK6
13	SA18	−IOW	SD10	DRQ6
14	SA17	−IOR	SD11	−DACK7
15	SA16	−DACK3	SD12	DRQ7
16	SA15	DRQ3	SD13	+5V
17	SA14	−DACK1	SD14	−MASTER
18	SA13	DRQ1	SD15	GND
19	SA12	−REFRESH	KEY	GND
20	SA11	CLK		
21	SA10	IRQ7		

PIN	J1/P1 (ROW A)	J1/P1 (ROW B)	J2/P2 (ROW C)	J2/P2 (ROW D)
22	SA9	IRQ6		
23	SA8	IRQ5		
24	SA7	IRQ4		
25	SA6	IRQ3		
26	SA5	–DACK2		
27	SA4	TC		
28	SA3	BALE		
29	SA2	+5V		
30	SA1	OSC		
31	SA0	GND		
32	GND	GND		

Note that the J2/P2 connector starts numbering the pins with 0. This can cause a problem for some PCB layout packages that expect the first pin of a device to be pin 1.

STD Bus

Much older than the PC/104, the STD bus has been used in a large number of embedded systems. Originally based on timing signals from the Zilog Z-80 microprocessor, the STD bus is available in 8- and 16-bit versions. The bus is based on a 56-pin edge connector, which originally supported a 64 K (16-bit) address space and an 8-bit data bus, so going to wider buses with more memory addressing capability has required multiplexing some of the pins. The eight upper address lines are multiplexed with the lower 8 data bits to provide 24 address bits. If a 16-bit data bus is used, the upper 8 data bits are multiplexed with the upper eight address lines. The STD bus pinout follows:

Component Side		Solder Side	
Pin	Signal	Pin	Signal
1	Vcc	2	Vcc
3	GND	4	GND
5	Vbb1	6	Vbb2
7	D3/A19	8	D7/A23
9	D2/A18	10	D6/A22
11	D1/A17	12	D5/A21
13	D0/A16	14	D4/A20
15	A7	16	A15/D15

Component Side		Solder Side	
Pin	Signal	Pin	Signal
17	A6	18	A14/D14
19	A5	20	A13/D13
21	A4	22	A12/D12
23	A3	24	A11/D11
25	A2	26	A10/D10
27	A1	28	A9/D9
29	A0	30	A8/D8
31	WR* (write strobe)	32	RD* (read strobe)
33	IORQ* (I/O sel)	34	MEMRQ* (memory sel)
35	IOEXP* (I/O expansion)	36	MEMEX* (memory exp)
37	RFSH* (refresh)	38	MCSYNC*
39	STATUS 1	40	STATUS 2
41	BUSACK* (bus ack)	42	BUSRQ* (bus request)
43	INTAK* (interrupt ack)	44	INTRQ* (interrupt req)
45	WAITRQ* (wait request)	46	NMIRQ* (NMI interrupt)
47	SYSRESET*	48	PBRESET*
49	CLOCK*	50	CNTRL*
51	PCO* (priority chain out)	52	PCI* (priority chain in)
53	AUX GND	54	AUX GND
55	AUX + V (+12 V)	56	AUX − V (−12 V)

Note: Signal names separated by slash (/) are multiplexed pins with two functions.

An STD bus system consists of a passive backplane with (typically) 4 to 20 slots, a plug-in CPU, and peripheral boards. The STD bus originally was used mostly with proprietary (non-PC) CPU designs. As the PC architecture became more attractive, STD bus boards and systems became available with the same architecture as a PC and able to run DOS or Windows. The number and type of peripheral boards (timers, I/O controllers, standard interfaces, data conversion, etc.) available for the STD bus is about the same as for the ISA or PC/104 buses.

Figure 9.2 shows the outline of the STD bus boards, which are about $4\frac{1}{2} \times 6\frac{1}{2}$ inches in size.

There is a newer version of the STD bus, STD-32, which supports 8-, 16-, and 32-bit transfers and a 32-bit address space. STD-32 uses interleaved connectors, and a STD-32 backplane will support older STD cards, allowing a mix of 8- and 32-bit cards in a system.

VME Bus

The VME bus was based on the Motorola 68000 signals. Using 96-pin DIN (a European standard) connectors, the backplane may be one to three

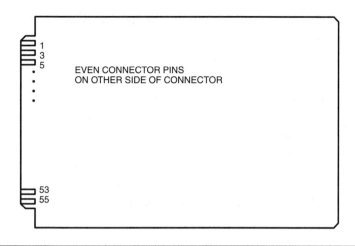

1
3
5

EVEN CONNECTOR PINS
ON OTHER SIDE OF CONNECTOR

53
55

Figure 9.2
STD Board Outline.

connectors wide and up to 20 or so slots long. The VME bus supports daisy-chained interrupts. It normally is associated with larger and costlier systems.

VME boards come in two sizes: 3U and 6U. Both are approximately 6.3 inches (160 mm) deep, although there is a longer version used by some systems. 3U boards have a single 96-pin VME connector and are about 3.9 inches (100 mm) wide. 6U boards have two connectors and are about 9.2 inches (233 mm) wide. There also is a three-panel-wide 9U board used in some systems; the third connector is user defined.

CompactPCI

A drawback to the standard ISA bus in a PC (and the similar PC/104 bus) is speed. ISA is limited to 16-bit transfers and, for compatibility reasons, limited in speed. The PCI bus in a PC overcomes some of these limitations, with a high-speed bus that supports 64-bit transfers and has a more flexible interrupt structure. The original 33 MHz PCI bus supports burst transfer rates of 133 MB/sec using a 32-bit mode, and 266 MB/sec using a 64-bit mode. However, the PCI bus, as implemented in a PC, still has drawbacks for industrial applications, since it uses edge connectors and a single-screw hold-down mechanism similar to the ISA.

The CompactPCI adapts the PCI bus to industrial and embedded applications. Like VME cards, CompactPCI boards are based on the Eurocard industry standard. CompactPCI boards come in 3U and 6U sizes. The connector is a 5 row × 44 pin connector, with 2 mm pin-to-pin spacing. The cards are held in place by a rail attached to the card-cage frame at each end with screws for secure mounting.

CPU on a Module

It is possible to purchase a "PC" in a module. ZF Microsystems (Palo Alto, CA), for instance, makes 386-, 486-, and 586-based modules that mount like a large surface mount device. Their OEMmodule 486 includes a 100 MHz 486SX, 2 MB of flash memory, 2 MB RAM, and standard PC peripherals such as IDE and floppy interfaces, a printer port, and two serial interfaces. The module is only 2.2 × 3 inches. The ZF Microsystems Megaton is a 586-based module in a ball grid array package. It includes:

PCI bridge interface

USB and I²C interfaces

IEEE-1284-compatible parallel port

Keyboard and mouse interfaces

Floppy and IDE hard drive controllers

Real-time clock

Two 16550-compatible UARTs

PWM generator

Watchdog timer

Decode logic for external flash memory

ZF Microsystems sells PC/104-based boards using these modules but also sells the modules separately. You can incorporate the modules into your design, giving you a custom design while retaining the advantages of the PC architecture.

CPU on a Chip

The AMD Élan SC520 Microcontroller provides a 32-bit, 100 MHz, 586 CPU core with several integrated peripherals. These include:

Integrated PCI host bridge

SDRAM controller

Programmable Interrupt Controller

PC-compatible timer

PC-compatible DMA controller

Two 16550-compatible UARTs

Real-time clock with battery backup

Three general-purpose timers

Watchdog timer

Synchronous serial interface

Programmable address decoding (chip selects)

32 general-purpose I/O pins

The Élan SC520 is optimized for embedded applications and provides a highly integrated solution when a PC-compatible embedded controller is needed.

Chapter 10 will examine some advanced microprocessor concepts.

Advanced Microprocessor Concepts   *10*

This chapter provides an overview of some features that are used to improve processor performance or to solve certain design problems.

Combination ICs

Most microprocessor designs that use external memory require both ROM/PROM and SRAM. Many manufacturers produce ICs that combine both flash ROM and SRAM in a single package. The Toshiba TH50VSF0302 is one such part, combining $1 M \times 8$ flash with $128 K \times 8$ SRAM. The Toshiba part is designed for 2.7–3.3 V operation, comes in a 48-pin ball grid array (BGA) package, and is available with either a top or bottom boot block (see Chapter 2). The SRAM and flash share a common data and address bus and the device has an access time of 100 ns.

Toshiba and other manufacturers make other combination parts that include bus widths up to 16 bits and memory densities to 2 MB (flash) and 512 K (SRAM).

Waferscale Integration (WSi) takes the combination chip concept a step further. Its PSD813F family of parts interfaces to most 8-bit processors. The PSD813F contains 128 K of flash ROM, 2 K of SRAM (optional, not on all versions), a decode PLD, a CPLD, and 27 I/O pins.

The decode PLD is used to decode the flash ROM, RAM, and other peripherals. It also can generate chip select outputs for other devices in the system. The CPLD can be used to implement general-purpose logic, including counters. The I/O pins can be used as output ports from the microprocessor, outputs from the PLD, or latched address outputs.

As mentioned, the PSD813F family has multiple parts, including some that include an OTP memory, and 256 K EEPROM.

Pipeline (Prefetch) Queue

To speed execution, some processors implement a pipeline, sometimes called a *prefetch queue*. This is because many CPU instructions are fairly complex, taking many clock cycles to perform. Multiply and divide instructions are good examples. While the processor is executing multiple-clock instructions, the bus normally is idle. In a processor with a pipeline, the bus logic goes ahead and gets the next few instructions in preparation for execution.

The Intel 80186/188 implements a pipeline by having the execution unit (EU) separate from the bus interface unit (BIU). While the EU is executing instructions, the BIU continues to fetch new instructions until the queue is full. If the next instruction in the pipeline happens to be one that can be executed very quickly, the one following already is in the pipeline and need not be fetched from memory.

A pipeline architecture keeps the CPU execution speed from being bogged down by slow memory. While the CPU is executing multiple-clock instructions, the pipeline uses those clock cycles to fill up with instructions. However, the *average* rate of instruction execution cannot exceed the memory bandwidth or the pipeline never will get ahead of the CPU and so provides no advantage.

The Motorola Coldfire CPU series takes the pipeline concept further. A drawback to a pipeline architecture is that, if a branch instruction is executed, all the prefetched instructions must be discarded and the pipeline refilled from the new address.

The MCF5307 is a 32-bit, Coldfire-family processor that fetches and partially decodes the instructions in the pipeline. If the decoding logic detects certain branch instructions, the pipeline will begin fetching instructions from the *new* address, in anticipation of the branch being taken. If the branch is conditional and *not* taken, then the new instructions are discarded and prefetching resumes from the addresses following the branch instruction.

Of course, this type of decoding has limitations. Suppose that a branch instruction uses an indirect address, contained in a register, and the register contents depend on an instruction still in the pipeline. Obviously, the pipeline logic, for any processor, cannot prefetch data since the destination address is not known.

Interleaving

Interleaving is used to allow a fast CPU to access slower memory without wait states. Figure 10.1 shows a simple timing diagram that illustrates the concept

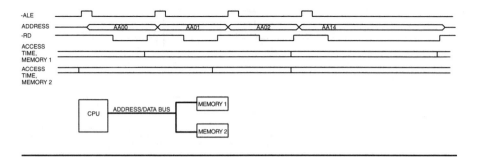

Figure 10.1
Interleaving.

of interleaving. In this example, an Intel-type bus was chosen because the ALE signal provides a reference for the processor cycles.

Two memories are shown in the figure. Each has an access time longer than the bus cycle time. Ordinarily, this would require the insertion of wait states. However, if each memory is accessed on every *other* cycle, the two memories together can keep up with the CPU. Each memory access starts in a cycle when the other memory is being read. In the figure, Memory 1 is accessed on every even-numbered address and Memory 2 on odd-numbered addresses.

Interleaving works only as long as the processor executes sequential address cycles. The access time for one memory device starts in the bus cycle for the other device, so the next address for each device must be predictable. In the example shown, the CPU is accessing a hex address of AA00 then AA01 then AA02 (these are just arbitrary addresses chosen for this example). After reading location AA02, the processor jumps to AA14. This memory access cannot be interleaved since the new address could not be predicted, so wait states must be inserted so that the memory can catch up with the CPU. You can see this in the figure, where the access to AA14 is longer than the preceding bus cycles.

A form of interleaving is performed in many microprocessor designs when interfacing to slower peripherals. Figure 10.2 shows a microprocessor connected to an analog-to-digital converter (ADC). When the microprocessor wants to read the ADC, it could start the ADC conversion, then wait until the conversion is complete. But this would waste time while the CPU is polling the ADC. Instead, the CPU starts a conversion, then goes away and does other things. At some regular interval, the CPU reads the ADC result and starts the next conversion. This technique can be applied to a number of different peripheral types. Two ADCs could be interleaved in the same way as memory accesses, permitting the conversions to overlap.

Advanced Microprocessor Concepts

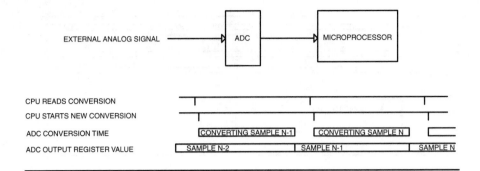

Figure 10.2
ADC Interleaving.

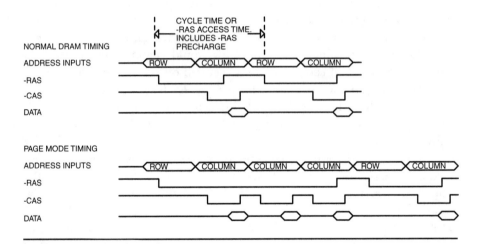

Figure 10.3
Burst Mode DRAM Access.

DRAM Burst Mode

Many dynamic memories have some form of burst mode of operation that permits faster access. Figure 10.3 shows how burst mode operation compares to the normal mode of operation in a dynamic RAM (DRAM). In normal operation, each cycle is initiated by -RAS, followed by -CAS. The access time of the DRAM is the -RAS access time, and the fastest rate the device can be accessed is the random access cycle time (a parameter you will find on the DRAM data sheet).

Figure 10.3 also shows *page mode*, which is the simplest type of DRAM burst operation. In this case, the -RAS signal goes low to latch the row address, but it stays low. Subsequent locations are read by strobing -CAS to latch a new column address. The -CAS access time is faster than the -RAS access time, so subsequent bytes can be read much more quickly than the first location. Any location in the selected row can be accessed in this way.

As soon as the CPU needs information from a different row, the -RAS line must be cycled and a new row address loaded. The access time for the first read from the new row is the -RAS access time, but subsequent reads from that row can be performed using burst mode access. A memory with a 100 ns -RAS access time typically would have a -CAS burst access time of around 60 ns. To take advantage of burst mode, the address decoding hardware must detect when the address changes to a different row (because the address bits from the CPU that make up the row address change). The -RAS signal must be cycled with the new row address, the first memory access is governed by the -RAS access time, and so the first bus cycle from the new row must be extended with wait states.

There are other enhancements to the page mode of operation, such as a fast page mode and extended data output (EDO). These all enhance performance by changing the burst mode timing, essentially making the -CAS access time shorter so that successive burst cycles are faster.

SDRAM

Synchronous DRAM (SDRAM) is a new type of DRAM that is optimized for high-speed microprocessors such as 586 and Pentium-class CPUs. SDRAM is a DRAM, and so it must be refreshed to retain the memory contents. However, synchronous DRAM operates at higher speeds than traditional DRAM. The most important difference is that SDRAM is synchronized to the CPU using a clock signal.

A typical SDRAM is the Toshiba TC59SM716/08/04. This is a 128 MB RAM, available as 32 MB × 4 bits, 16 MB × 8 bits, or 8 MB × 16 bits. The TC59SM716 comes in a 54-pin surface mount (TSOP) package, operates at 3.3 V, and is capable of transferring up to 133 megawords/sec. The signals on this SDRAM IC are as follows:

Data lines (16)

-CAS

-RAS

-WE

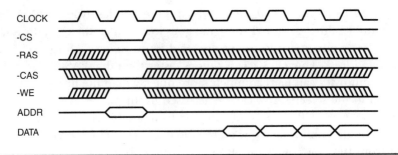

Figure 10.4
SDRAM Timing.

-CS

Clock

DQM (data bus select)

Bank select

Address signals

SDRAM ICs have -RAS, -CAS, and -WE signals like normal DRAM ICs. However, these signals have a different meaning on SDRAM. In addition, SDRAM has clock, a chip select (CS), bank select (BS) signals, and data bus select signals. Finally, the address lines on an SDRAM are used both to address the device and to select certain parameters.

Figure 10.4 shows the basic timing of a SDRAM read cycle. Note that all the input signals are synchronized to the rising edge of the clock signal. In the waveform shown, the CPU has requested a burst read of multiple words. The command is issued on one clock edge, and three clocks later, the data are available at the SDRAM outputs. Once the first word has been read, subsequent words are read on each clock cycle. Although not shown in the figure, accessing an SDRAM requires a -RAS cycle (also synchronous) to load the row address and activate the row.

Like an ordinary DRAM, the SDRAM uses a burst mode to read subsequent locations. In the case of SDRAM, a new location is read on each clock. The burst length is set with a Mode Register Set command. When this command is issued, the address bits are redefined as command bits. The meaning of the bits is as follows:

A0–A2: Burst length

A3: Addressing mode (sequential or interleaved)

A4–A6: -CAS latency

A9: Write mode

The -CAS latency tells the DRAM how many clock cycles (two or three) should elapse between a command being issued and data being available. This allows the DRAM delay to be set so that it matches the CPU clock. A fast CPU would select a three clock-cycle latency, a slower CPU (with a corresponding slower clock signal to the DRAM) would select two clock cycles.

The -RAS, -CAS, and -WE signals select the command mode. A partial list of these commands is as follows:

-RAS	-CAS	-WE	Command
0	0	0	Mode register set
0	0	1	Auto refresh/self-refresh entry/self-refresh exit
0	1	0	Bank precharge/precharge all
0	1	1	Bank activate
1	0	0	Write/write with auto precharge
1	0	1	Read/read with auto precharge
1	1	0	Burst stop
1	1	1	No operation

As you can see, there is more than one interpretation for each command state. Which command is executed depends on the state of an address line and what state the SDRAM already is in.

An SDRAM IC has 16 data lines. The data can be accessed in 8- or 16-bit words; the DQM signals determine which bytes are read. DQM also functions as a mask when writing, allowing either or both bytes of the pair to be written. This permits a word-wide processor to perform byte-oriented operations on the device. Of course, the DQM signals on multiple devices can be manipulated so that a 32- or 64-bit-wide memory array can be accessed as bytes, 16-bit words, or 32-bit words.

An SDRAM data sheet consists of 50 or so pages of timing diagrams and tables. Due to the high clock rates (66–125 MHz), SDRAM timing usually is accomplished with fast programmable logic or custom ICs. One advantage to SDRAM is the synchronous nature of the interface. Traditional DRAM requires delay lines or other timing devices to get the -RAS and -CAS strobes right. SDRAM synchronizes everything to the clock signal, which is a convenience since the control logic usually is synchronous anyway.

Some microprocessors, such as the AMD ElanSC520 microcontroller, include an SDRAM interface on-chip.

High-Speed, High-Integration Processors and Multiple Buses

Although the interfacing techniques introduced in Chapter 2 apply across all speed ranges, some special considerations are in order for interfaces to very fast processors. The AMD Élan SC520 is one example. The SC520 integrates a 586 CPU core with a number of peripheral functions. One is a fast interface to external flash memory. The 586 flash memory interface can run at 33 MHz, performing one fetch from the flash every 30 ns. Since most flash memories cannot operate at this speed, the CPU needs wait states to access the flash. It might seem reasonable to simply run the CPU at a slower clock and avoid wait states. However, the SC520 has other integrated interfaces, including a SDRAM interface. Operating the flash with wait states allows the SDRAM to run at full speed. In many cases, when using a PC-compatible processor like the SC520, the flash is used only when starting the system; normal operating code is stored in RAM.

Figure 10.5 is a block diagram of the Intel i960 VH processor. The i960 is a high-performance microprocessor family. The VH version has two external buses—a local memory bus and a PCI bus. The PCI (peripheral component interconnect) bus is a standard interface bus in the IBM-PC world. The i960 VH incorporates a PCI controller on-chip. The i960 also has a local memory bus for accessing DRAM or flash memory. The i960 VH has an internal 32-bit address space; the PCI bus can be made part of this address space or the VH address space can be independent of the PCI bus. Integration of the PCI bus onto the chip provides a very high level of performance on a standard interface.

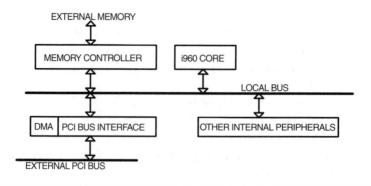

Figure 10.5
Intel i960 VH.

Cache Memory

One problem that occurs as processors get faster and faster is the bottleneck of accessing memory. On-chip speeds inside the CPU always are faster than the speed of external buses. For example, the PC-standard PCI bus at 66 MHz usually is driven by a CPU with a much faster internal clock. A 100 MHz PCI typically is connected to a 300 MHz or faster CPU. And, 100 MHz SDRAMs connect to 350 or 400 MHz CPUs.

The reason for this is that the logic delays inside the CPU are more controllable and more repeatable than those going off-chip. Also, signal paths inside the chip are only tiny fractions of an inch, versus longer traces on a PC board. This affects both the propagation delay and the transmission-line characteristics of the traces.

The bottom line is that a very fast CPU may be unable to execute instructions at full speed because it is starved for data from a memory that cannot keep up. One solution to this problem is the addition of cache memory. *Cache memory* is a fast memory located close to the CPU and operating closer to CPU speeds. Cache memory usually is implemented with very fast static RAM.

Cache memory is managed by a cache controller that fetches data from the main memory and stores it in the cache. Cache memory works because most microprocessor programs are repetitive in nature—the code loops around and around, executing the same string of instructions for some time before moving on to some other piece of code. When the CPU wants to execute code not in the cache, the cache controller gets the code from main memory (DRAM, usually) and moves it into the cache. Once in the cache, the code executes very quickly.

If cache is so fast, why not just make all the memory cache? The first reason is cost—building all main memory out of the superfast cache SRAM would make the memory prohibitively expensive. Second, cache SRAM ICs are larger than equivalent DRAM due to the larger cell size and added number of pins required. So, making the entire main memory out of cache parts would make the memory array physically larger, which would limit speed due to trace lengths.

Many CPUs, such as Pentium-class processors, go a step further, integrating a small cache onto the CPU chip itself. This provides a very fast cache memory, capable of keeping up with the CPU at full speed. However, since SRAM takes a significant amount of real estate on the CPU die, on-chip cache memory typically is smaller than off-chip cache memory. Many designs include both types of cache memory for maximum performance.

Processors with Multiple Clock Inputs and PLLs

Many microprocessors need more than one clock input. The AMD SC520 is an example of this. The SC520 requires two crystals (or external oscillators). One crystal runs at 32.768 kHz and provides a signal to the real-time clock and SDRAM refresh logic. The SC520 also has a 33 MHz input, which provides clocks to the CPU, PCI, and other internal peripherals.

As processor speeds exceed 30 MHz or so, it is difficult to get crystals to run the CPU. Fundamental mode crystals typically are unavailable above 30 MHz. The SC520, in addition to the clocks mentioned, requires 66 MHz for the SDRAM logic and 18.432 MHz for the UARTs. Clocks like this usually are generated by a *phase-locked loop* (PLL) inside the microprocessor IC.

While the complexities of PLL theory are beyond the scope of this book, a PLL can be thought of as a block of components that multiply a clock by some integer. Figure 10.6 shows a simplified block diagram of a PLL and a brief description of how the circuit works.

A microprocessor may contain multiple PLLs to generate more than one frequency. The SC520 has a PLL that generates 1.1882 (for the programmable timers) and 18.432 MHz (for the UARTs) from the 32.768 kHz input. Another PLL produces 66 MHz for the SDRAM interface from the 33 MHz input. The CPU core has a PLL that multiplies the 33 MHz input crystal by 3 or 4 to produce a 100 MHz or 133 MHz CPU clock.

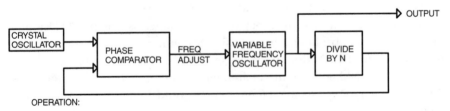

OPERATION:

PHASE COMPARATOR ADJUSTS VFO FREQUENCY SO THAT OUTPUT OF DIVIDER MATCHES CRYSTAL OSCILLATOR.

FOR DIVIDER OUTPUT TO MATCH OSCILLATOR OUTPUT, VFO FREQUENCY MUST BE OSCILLATOR FREQUENCY x THE DIVIDE VALUE (N).

EFFECT OF PLL IS TO MULTIPLY CRYSTAL OSCILLATOR FREQUENCY BY N.

Figure 10.6
PLL Block Diagram.

Embedded Microprocessor Systems

Multiple-Instruction Fetch and Decode

With the addition of on-chip cache memory to some microprocessors, a secondary performance improvement is possible. It is possible to build a microprocessor with a 32-bit interface to external memory, but a 64- or 128-bit interface between the internal cache memory and the internal CPU core. Figure 10.7 shows this arrangement.

Since the internal bus that interfaces the cache memory to the CPU is wider than the CPU word, it is possible to transfer multiple instructions to the CPU at once. With parallel hardware, the CPU can decode more than one instruction at a time, resulting in a very high level of performance. The Intel i960 does this, as does the Motorola Power PC. Of course, this greatly increases CPU complexity, as there must be some degree of parallelism and the instructions must be synchronized.

Clock-Synchronized Buses

Figure 10.8A shows a typical microprocessor bus read cycle. This is an Intel-type bus, such as in the 80186. The CPU generates a stable address, generates ALE to allow the external hardware to latch the address, and generates an -RD strobe to direct the external peripheral to place data on the data bus.

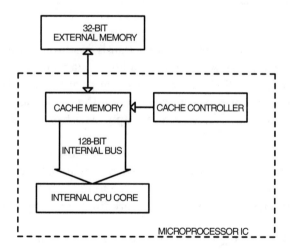

Figure 10.7
Wide Cache Memory.

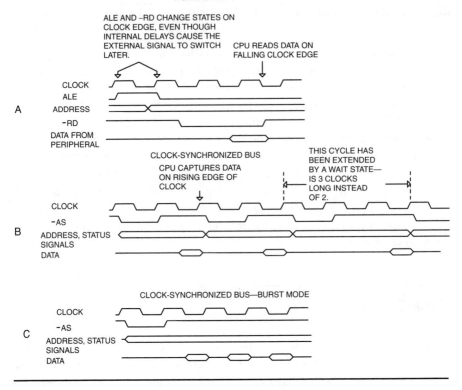

Figure 10.8
Clock-Synchronized Bus.

All microprocessors use internal logic that is synchronized to the internal clock. In the figure, -RD, ALE, and the address/status lines all are synchronized to the processor clock. Even though the peripheral enables data with the -RD strobe, the processor captures data on the clock edge. If this were a write cycle instead of a read cycle, the output data from the CPU would be synchronized to the clock.

Knowing this, it is possible to eliminate the -RD and -WR signals (-DS on a Motorola-style bus) and just synchronize everything to the processor clock. As long as the peripheral knows what clock edge to use and meets the setup and hold-time requirements, everything will work the same as if the strobe signals were there.

Figure 10.8B shows such a clock-synchronized bus. This example is based on the timing for the Intel i960 microprocessor. The address and status signals are driven onto the bus and the -ADS signal indicates a valid address. The address and status signals include all the status signals including the address

Embedded Microprocessor Systems

lines, DMA indication, -LOCK signal, and read/write signal. Essentially, everything the peripheral device needs to determine what kind of bus cycle is being started is available while -ADS is low.

The peripheral decodes the address and status signals, capturing them on the rising clock edge that occurs while -ADS is low. Before the next rising clock edge (-ADS has gone high), the peripheral places data on the data bus and the CPU captures it on the rising clock edge.

This scheme allows all the decoding logic to be synchronous. The catch is that the decoding logic or the peripheral must keep track of when the CPU expects data. The last cycle shown in Figure 10.8B illustrates a bus cycle that is extended by a wait state. The peripheral (or the decoding logic) must keep track of which clock it is on and drive the data lines on the right clock edge.

Processors that support such a clock-based bus often provide a burst mode of operation that is ideal for interfacing to burst-mode memories such as DRAM. The i960 supports such a burst mode, as shown in Figure 10.8C. The first cycle looks like the ones shown in Figure 10.8B, but subsequent cycles can transfer one word from memory per clock cycle (assuming appropriate memory speed). Although not shown in the figure, the i960 has two additional address signals that are cycled through when the burst mode is used, allowing up to four words to be accessed.

The i960 also supports a pipelined mode. In this mode, each data transfer takes two clock cycles, like the bus cycles in Figure 10.8B. However, the cycles overlap, so that, while the CPU is reading data for one address, it is placing the address and status information for the *next* bus cycle on the bus. This allows one word to be transferred per clock cycle, just as in burst mode. Obviously, the decoding logic must detect this condition, capturing the address/status information and enabling the right peripheral or memory at the right time.

Intel is not the only manufacturer whose processors use a clock-synchronized bus. The Motorola Power PC uses a bus that has different signal names but timing that is very similar to the i960. In fact, nearly all high-performance processors use a clock-synchronized bus of some type.

Interfacing memory and peripherals to a clock-synchronized microprocessor is similar to interfacing to an ordinary microprocessor. The same considerations apply for setup, access, and hold timings. The differences are that, first, the times typically are much shorter due to the higher clock rates. Second, the interface is synchronous, so normally you will use some type of PLD or ASIC. The interface logic has to keep track of the type of bus cycle (burst, pipelined, etc.) and insert wait states for peripherals that need them. When interfacing to a synchronized bus, you are likely to find that you need a wide variation in bus cycle times. You may use a fast DRAM or SDRAM that matches the CPU speed, needing no or one wait states, and a slower peripheral that needs three or four wait states. With a synchronized bus, the propagation delays in the decoding logic become significant, although that is somewhat alleviated by the synchronous nature of the design. Finally, to achieve maximum performance, the interface logic must recognize and support special features of the bus, such as the burst mode.

On-Chip Debug

The addition of on-chip cache memory and high-speed processors complicates debugging. If instructions are executed from the on-chip cache, there is no external indication on the processor pins of what is going on. Prefetching causes problems as well; an instruction may be fetched from memory but never executed. An in-circuit emulator could monitor execution of these instructions, but the high clock rates of current processors make such an emulator difficult to build.

Another problem with emulators for high-performance processors is packaging. In the early days of microprocessors, all ICs came in DIP packages that could be socketed easily. The microprocessor could be removed from the socket and an emulator installed. Today, many microprocessors come in surface-mount packages that cannot be socketed. Removing the chip from the board to install an emulator is not possible, even if there were a way to attach the emulator to the board.

To simplify debug of high-performance processors, many manufacturers include on-chip debugging resources. As mentioned in Chapter 6, the x86 family of processors, starting with the 386, includes on-chip debug registers. Figure 10.9 shows the configuration of the x86 debug registers for the Pentium processor.

The Pentium has eight debug registers, DR0 through DR7. All registers are 32 bits wide. DR4 and DR5 are reserved, so only six registers actually are used. DR0 through DR3 are linear breakpoint address registers, written with the address of the breakpoint. This is an unsegmented, 32-bit address (if you do not know what *unsegmented* means, do not worry about it; it is a function of the x86 architecture).

Register DR7 controls what type of breakpoint is executed. Each address register has two LEN and two R/W bits; the encoding of the LEN and R/W bits is shown in the figure.

The L0–L3 and G0–G3 bits individually enable the four breakpoints. L0–L3 are used for local breakpoints (cleared after a task switch) and G0–G3 are

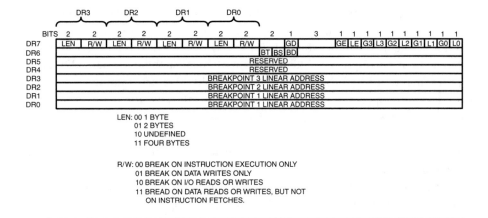

BITS	2	2	2	2	2	2	2	2	2	1	3	1	1	1	1	1	1	1	1	1	1
DR7	LEN	R/W	LEN	R/W	LEN	R/W	LEN	R/W			GD		GE	LE	G3	L3	G2	L2	G1	L1	G0	L0
DR6									BT	BS	BD										
DR5	RESERVED																				
DR4	RESERVED																				
DR3	BREAKPOINT 3 LINEAR ADDRESS																				
DR2	BREAKPOINT 2 LINEAR ADDRESS																				
DR1	BREAKPOINT 1 LINEAR ADDRESS																				
DR0	BREAKPOINT 1 LINEAR ADDRESS																				

DR3 DR2 DR1 DR0

LEN: 00 1 BYTE
01 2 BYTES
10 UNDEFINED
11 FOUR BYTES

R/W: 00 BREAK ON INSTRUCTION EXECUTION ONLY
01 BREAK ON DATA WRITES ONLY
10 BREAK ON I/O READS OR WRITES
11 BREAD ON DATA READS OR WRITES, BUT NOT
ON INSTRUCTION FETCHES.

Figure 10.9
Intel Pentium Debug Registers.

used for global breakpoints, which are not cleared after a task switch. This is needed because a task switch may put something else in the memory area pointed to by the address register, and the breakpoint would be invalid.

Debug features such as these permit a software debugger to simulate some of the features of an in-circuit emulator. A breakpoint can be executed if the processor writes to certain I/O addresses, for instance, or if a particular variable is accessed.

The x86-family are not the only processors with on-chip debug features. Most high performance 32- or 64-bit processors include some type of on-chip debug. Motorola uses a method called *background debugging mode* (BDM). BDM allows an external host PC (with appropriate software, of course) to monitor and control the target CPU. BDM uses three processor pins, a clock, data in, and data out. These pins perform more than one function, depending on the mode of the BDM interface. When transferring data, the BDM pins function similar to an SPI port. The BDM data word transferred to the PC is 17 bits long. BDM permits the user to read and write registers, read and write memory, and perform other basic debugging functions.

In the past, on-chip debugging resources were available only on 16- to 64-bit microprocessors, not on smaller microcontrollers. For many microcontrollers, the on-chip debugging circuitry would be a significant portion of the IC die. However, Microchip has started adding in-circuit debugging to the PIC processors. The PIC 16F877 has added on-chip circuitry that permits a breakpoint to be set and memory to be examined. Compared to the resources on a Pentium or Power PC, these may seem inadequate. But it is a big leap from where microcontroller debug was in the past. And remember, microcontrollers often are used in simple applications, where extensive debug support is not needed. A memory dump feature may be all that is required with only 256 bytes of on-chip RAM. The microchip debug capability is enabled by programming a bit when the microcontroller EPROM is programmed.

Many microprocessors implement a JTAG interface for debugging. The JTAG (joint test action group) interface is a standardized serial interface that permits automatic test equipment to serially read and write the contents of internal registers in the IC. The JTAG interface standard (IEEE 1149) is flexible enough that it also can support on-chip debugging capabilities.

The AMD SC520 uses the JTAG interface to provide debug support. An internal memory stores trace information about program execution. Of course, with a serial interface, there is no way to track every instruction in real time, so the trace information is partial. The software in the host PC has to do some of the work of decoding the debug information from the chip.

Memory Management Hardware

Many advanced microprocessors include hardware for memory management. The features provided by memory management include the following.

Memory Protection

As mentioned in Chapter 3, there is nothing to prevent a berserk program from writing all through memory. In a system with a memory management unit (MMU), each program is limited to its own area of memory and cannot corrupt memory allocated to other programs.

Write Protection

Using an MMU, certain areas of memory can be set aside as read-only, even though they are physically implemented as RAM. The MMU detects any attempts to write to those memory areas.

Relocation

A program may be written with absolute branch addresses and it may access absolute memory locations. Such programs cannot be relocated because the addresses would all be wrong. The MMU can translate the addresses, allowing the programs to be executed from any location in memory.

Supervisor

Processors that have a MMU also have multiple privilege levels. Supervisor will be one of these. The supervisor level allows the MMU to be programmed, among other things. Typically, programs that are not at the supervisor level cannot execute certain instructions, such as instructions that disable interrupts or modify the interrupt vector table.

As an example, we take an overview look at the memory management scheme used by Intel for the x86 family. The Intel memory management scheme is an outgrowth of the original 8086 segmentation architecture.

Segment Registers Segment registers were introduced with the 8086 to permit the 16-bit processor to access up to a megabyte of memory (which requires 20 address bits). The 16-bit segment register contents are shifted left four places and added to the 16-bit offset to make a 20-bit address. The memory thus is divided into 64 K segments. If a program wants access to two memory locations that are more than 64 K apart, two different values must be used in the segment register to do so. Similarly, if the program itself is bigger than 64 K, the segment register that points to the code area must be changed when the program rolls over or jumps into a section of code that cannot be reached with the current segment register and program counter. For example, if the code segment register contains C000 and the program counter contains FFFF, the current instruction will come from the absolute address CFFFF. You would expect the next instruction to come from D0000, but that is not what happens. Instead, the PC rolls over to 0 while the code segment stays the same, so the next instruction comes from C0000. The code segment register must be changed to reach anything above D0000.

The original 8086 provided four segment registers: code segment, data segment, stack segment, and extra segment.

With the introduction of the 386 processor, a new method was needed. The 386 is a 32-bit machine, with a 32-bit address bus. To accommodate this architecture, the segment registers in the 386 (and above) processors are 32 bits wide and point to a table of descriptors. When the CPU wants to access memory, the segment (now called a *selector*) register is used to obtain a 64-bit entry from the descriptor table. This entry contains:

- The absolute 32-bit start address of the segment
- The upper limit of the segment
- The status, privilege level, segment type, whether the segment is present, and the like

So a program can be loaded anywhere in memory; accesses to memory (including code, data, and stack) are translated into absolute 32-bit addresses using the descriptor table.

Privilege Levels   The Intel MMU provides for four privilege levels: Level 0 is the highest level and permits access to anything in the system, including the MMU itself and all instructions in the instruction set. The operating system kernel will be at level 0.

The other three levels, 1 through 3, have fewer privileges. The essential point is that the MMU will not permit any memory access that is off-limits to a program at a given privilege level. A memory segment can be set so that it is read-only for levels 1 and below. A program at privilege level 0 can write to that segment, but a program at level 1, 2, or 3 can only read it. Other registers in the MMU control things like what privilege level is permitted to disable interrupts or to modify the MMU registers.

Motorola

The Motorola memory management scheme on the 68060 is different from Intel's, but the result is the same—a table is used to translate a logical address to a physical address. The 68060 has seven key MMU registers. One register points to a descriptor table for the supervisor level and one register points to the user descriptor table. One register controls various functions like page size (4 K or 8 K), and four registers provide translation information for code and data (two registers each).

Exception Handling

What happens when a program tries to write to read-only memory or disable interrupts when its privilege level is not high enough? When this happens, an *exception* is generated by the MMU. An exception is similar to an interrupt and handled much the same way. Exceptions are not disabled by disabling interrupts, although the MMU can be programmed not to generate exceptions. The exception handler, part of the operating system, decides what to do if an illegal operation is attempted.

Appendix A:
Example System Specifications

System Definition

System Description

The system is a swimming pool timer that cycles the AC pump motor on a swimming pool.

The power input is 9–12 V DC from a wall-mount transformer.

The pump is a $^1/_2$ hp single-phase AC motor, controlled by mechanical relay. Relay is remote from control unit, located in weatherproof box near pool pump motor.

Provision is to be made for a switch closure input that prohibits pump operation if the water level is low.

The user can set the length of time the pump is on and the length of time it is off. An override is available to permit turning off the pump when it is on for maintenance and turning the pump on when it is off so that chemicals can be added.

On/off/override time is to be adjustable in 30-minute increments from one-half hour to 23 hours. A display will indicate the on/off condition of the pump, the time remaining, and if the pump is in the override mode. The display also will indicate the condition of the water low monitor.

A minimum number of switches/knobs will be used.

User Interface

Display   Four seven-segment digits: two digits for hours, two for minutes. Also three LEDs: SET, ON, and OVERRIDE. The LEDs are to be high intensity for daylight readability.

Keypad   There are four keys: SET, ON, OFF, FCN.

Operation   The display will indicate the time remaining before pump switches on or off.

After reset, ON time will be set to 8 hours, 30 minutes. Off time will be set to 8 hours, 0 minutes.

Display will flash to indicate that power has been removed.

After power-up, ON and OFF override will not be allowed until SET has been pressed by user. Pressing ON will activate ON override. Pump will be turned on for 30 minutes, the display will show the override time, and the override and ON LEDs will be lit. Each successive push of the button will increment the override time. Normal time will continue to count while in the override mode. When the override time expires, time keeping and display will revert to normal mode. Pressing OFF will activate the OFF override, with the same characteristics as the ON override.

Pressing OFF while in ON override or pressing ON while in OFF override will terminate override mode. Time keeping and display revert to normal mode.

ON override may be used while the pump is on normally to extend the ON time to up to 24 hours. Similarly, the OFF override may be used while the pump is off normally.

Setting Time

When user presses SET, the timer enters the time set mode. The set LED goes on. Pressing ON after SET will light the ON LED and show the current ON time, not the time remaining. Each press of ON will increment the time by 30 minutes, until the time reaches 24 hours, then the time will roll over to 30 minutes. Pressing SET terminates the time set mode and stores the time, and the SET LED goes off. OFF time setting works the same as ON time setting.

While setting ON time, pressing OFF will save the ON time and change to the OFF set mode. Similarly, pressing ON while setting OFF time will save the OFF time and switch to the ON time set mode.

Water Low

If a low water condition is detected by closure of the water low switch, the pump will turn off if it is on. If the pump is already off, it will not be permitted to turn on. Any time that the water low condition is detected and the pump should be on, the ON LED will flash to indicate the problem. The water low switch input will be filtered to prevent spurious transitions.

Example System Hardware Specifications

Initial Hardware Specification (Predesign)

Display:

Four seven-segment LED displays (hours, minutes)

ON LED (high intensity)

SET LED (high intensity)

OVERRIDE LED (high intensity)

Keys:

SET: Enables time set

ON: On override, on time set

OFF: Off override, off time set

FCN: Undefined

Other inputs:

Water low switch closure

Power: 9–12 volts DC input, using coaxial connector. Onboard 5 V regulator. Polarity protected.

Outputs:

Relay on/off. Relay powered by unregulated DC input.

Other outputs:

Watchdog timer required.

Circuit Description (Postdesign)

CPU: 8031, 6 MHz input clock.

EPROM: 8 K × 8, external (2764). No internal ROM

8031 port usage:

Ports 0,2: Address/data bus for external memory access
Port 1: LED/display control
   Bit 0: Zero enables minutes, ones display digit
   Bit 1: Zero enables minutes, tens display digit
   Bit 2: Zero enables hours, ones display digit
   Bit 3: Zero enables hours, tens display digit
   Bit 4: One turns on override LED
   Bit 5: Unused
   Bit 6: One turns on set LED
   Bit 7: One turns on ON LED
Port 3:
   Bit 0: Unused
   Bit 1: 1 turns on motor relay
   Bit 2: Toggle to trigger watchdog timer
   Bits 3–5: Unused
   Bits 6, 7: External register access (RD/WR)

External registers: One read buffer, one write register. No address decoding—read from any external data memory address will enable the read buffer and any external write will clock the write register.

External read buffer:

D0, D1: Unused, read as 0.
D2: 0 = FCN key pressed
D3: 0 = OFF key pressed
D4: 0 = ON key pressed
D5: 0 = SET key pressed
D6: Unused
D7: 0 = Water Low switch closed

External write register: LED segments, writing 1 turns on segment.

D0: Segment A
D1: Segment B
D2: Segment C
D3: Segment D
D4: Segment E
D5: Segment F
D6: Segment G
D7: Decimal point

LED segment definition:

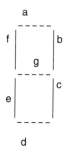

LEDs are not decoded—software directly writes LED segments. Numeric to seven-segment decoding must be performed in software.

Software must multiplex (scan) display digits.

Switch inputs are not debounced.

Watchdog timer has approximately a 0.5 second timeout.

Example System Software Description

Requirements

Implement functionality as described in system definition.

Implement additional functionality as described in hardware definition.

CPU Resource Usage

Timer 1: 250 Hz interrupt

Timers 0 and 2: Unused

Ports: As decribed in hardware definition.

Bit 3.4 is reserved as diagnostic output for oscilloscope.

Functional Software Description for Pool Pump Timer

This is a high-level logical description, one step above pseudocode.
Reset logic:

```
Turn all display digits off.
Set mode to powerup.
Clear all variables.
Set ON time to 8:30.
Set OFF time to 8:00.
Set current time to ON time. This will turn the pump on.
Start of background loop:
    If counting ON time or if in ON override,
     If water level OK, turn pump on.
    If counting OFF time, or if in OFF override
    or if water level low, turn pump off.
    If time rolled over from ON to OFF,
        Switch to counting ON time.
        Set current time to OFF time.
If time rolled over from OFF to ON,
    Switch to counting ON time.
    Set current time to ON time.
    If mode is powerfail,
     If set pushbutton pressed, set mode to normal timekeeping.
    If mode is normal timekeeping,
     If ON push button pressed (ON override)
      If override time = 0:0 (first button press),
        Set to ON override mode
        Set override time to 0:30.
     If Override time was 0:0,
      If in ON override,
       add 30 to override time
       If override time = 24:0, set override time to 0:0.
     If in OFF override (ON pressed while in OFF override),
        Set override time to 0:0 (exit override).
     If OFF push button pressed (OFF override)
      If override time = 0:0 (first button press),
```

Set to OFF override mode
Set override time to 0:30.
If Override time was 0:0,
   If in OFF override,
      add 30 to override time.
   If override time = 24:00, set override time to 0:0.
   If in ON override (OFF pressed while in ON override),
      Set override time to 0:0 (exit override).
If SET push button pressed,
   Set mode to timeset
   Display ON time
   Set override time to 0:0.
If mode is time set,
   If SET push button pressed,
      Set mode normal timing.
      If we were setting ON time, set ON time to displayed time.
      If we were setting OFF time set OFF time to displayed time.
If ON button pressed,
   If setting ON time, increment displayed time.
   If setting OFF time,
      Set OFF time to displayed time
      Display ON time.
If OFF button pressed
   If setting OFF time, increment displayed time.
   If setting ON time,
      Set ON time to displayed time
      Display OFF time.

End of background loop.

Example System Software Pseudocode

Variables for Background Loop

MODE: Operating mode.

0 = POWERFAIL
1 = timing
ONOFF = 1, ON timing
ONOFF = 0, OFF timing
2 = set time

MINUTE: current time, minutes count

HOUR: current time, hours count

ONMIN: ON time minutes

ONHOUR: ON time, hours

OFFMIN: OFF time, minutes

OFFHOUR: OFF time, hour

OVMIN: override, minutes

OVHOUR: override, hours

PRMIN: set time, minutes

PRHOUR: set time, hours

SEMODE: 1 = set on time, 0 = set off time.

ONOFF: 1 = on timing, 0 = off timing. Mode 1 only.

VOFLAG: 1 = on override, 0 = off override,

   valid only in override mode.

Variables That Pass Information Between the 250 Hz Timer and Background Code

Switch flags are set when the switch is debounced and cleared by the background code when processed. The timer will not set the flag again until the push button (PB) is released and pressed again. If flag is not cleared by the background code, the timer will clear the flag when the PB is released.

ONFLAG: ON switch pressed

OFFLAG: OFF switch pressed

FCFLAG: FCN switch pressed

SEFLAG: SET mode switch pressed

MTFLAG: low water detected

TFLAG: 1 = time rolled over

Variables Used by 1/250 Hz Timer

DBCOUN: push button debounce counter

HUND: 1/250 second counter

SECOND: current seconds count

MTCOUN: low water debounce counter

DISPLY: which display digit is on, 0–3

BLFLAG: blinking flag for display

Reset Processing

Turn all displays off.

Set MODE = 0 (powerup mode).

Initialize variables to 0.

Set ON time to 8:30 (ONHOUR = 8, ONMIN = 30).

Set OFF time to 8:00 (OFFHOUR = 8, OFFMIN = 30).

Set current time to ON time. (HOUR = ONHOUR, MINUTE = ONMIN, ONOFF = 1).

Background Loop

```
If ONOFF set (ON timing),
OR if in override mode and VOFLAG set (ON override mode),
   If MTFLAG = 0 (water level OK), Turn pump on.
If not ONOFF (Off timing),
OR if override time 0 and VOFLAG not set (OFF override),
   Turn pump off.
If TFLAG (time rolled over),
   Clear TFLAG.
   If ONOFF (ON timing, need to change to OFF timing),
   Clear ONOFF
   HOUR = OFFHOUR
   MINUTE = OFFMIN (current time = OFF time).
   Else (ONOFF was not set, OFF time, change to ON time),
   Set ONOFF
   HOUR = ONHOUR
   MINUTE = ONMIN (current time = ON time).
If powerfail occurred, switch to normal timing only if
SET button pressed.
If MODE = 0 (powerfail)
   If SEFLAG (SET PB pressed),
   Clear SEFLAG
   Set MODE = 1 (normal timing).
If MODE = 1 (normal timing),
If ONFLAG set (ON PB pressed),
   Clear ONFLAG.
   If override time = 0:0 (OVMIN = OVHOUR = 0),
   (User has selected ON override.)
   Set VOFLAG
   Set OVMIN to 30.
   Else (override time 0:0, user has pressed ON
      while in override),
   If VOFLAG (ON pressed in OFF override, cancel
         override),
   Set OVMIN = OVHOUR = 0:0 (override time = 0:0)
   Else (ON pressed while in ON override, incr time),
   Add 30 to override time
   If override time = 24:00, set override time to 0.
If OFFLAG set (OFF PB pressed),
```

Clear OFFLAG.
If override time = 0:0 (OVMIN = OVHOUR = 0),
(User has selected OFF override.)
Clear VOFLAG
Set OVMIN to 30.
Else (override time >> 0:0, user has pressed OFF
  while in override),
If not VOFLAG (OFF pressed in ON override, cancel
              override),
Set OVMIN = OVHOUR = 0:0 (Override time = 0:0)
Else (OFF pressed while in OFF override, incr time),
Add 30 to override time.
If override time = 24:00, set override time to 0.
If SEFLAG (SET PB pressed),
Set MODE = 2 (time Set)
Set OVMIN = OVHOUR = 0:0 (Override time = 0:0)
Set PRHOUR = ONHOUR
Set PRMIN = ONMIN (display ON time)
Set SEMODE = 1 (ON time set).
If MODE = 2 (time set),
    If SEFLAG (SET PB pressed, exit time set),
    Clear SEFLAG.
    If SEMODE = 1 (ON time set),
    ONHOUR = PRHOUR
    ONMIN = PRMIN (Store displayed time as ON time)
    MODE = 1.
    Else (SEMODE = 0, OFF time set),
    OFFHOUR = PRHOUR
    OFFMIN = PRMIN (OFF time = displayed time)
    Mode = 1.
If ONFLAG (ON PB pressed while in timeset mode),
    Clear ONFLAG.
    If SEMODE = 1 (ON pressed in ON timeset, incr display time),
    Add 30 to displayed time.
    If time = 24:00, set displayed time to 0:30.
    Else (SEMODE = 0, ON pressed in OFF set, save OFF time),
    OFFHOUR = PRHOUR
    OFFMIN = PRMIN (OFF time = displayed time)
    SEMODE = 1
    PRHOUR = ONHOUR
    PRMIN = ONMIN (displayed time = ON time).
If OFFLAG (OFF PB pressed while in timeset mode),
    Clear OFFLAG.
    If SEMODE = 0 (OFF pressed in OFF timeset,
            increment display time),

Add 30 to displayed time.
If time = 24:00, set displayed time to 0:30.
Else (SEMODE = 1, OFF pressed in ON set, save ON time),
ONHOUR = PRHOUR
ONMIN = PRMIN (ON time = displayed time)
SEMODE = 0
PRHOUR = OFFHOUR
PRMIN = OFFMIN (displayed time = OFF time).

End of background loop.

Timer Interrupt Logic

Trigger watchdog timer.
Increment HUND.
If HUND = 125 (0.5 sec rollover), toggle BLFLAG.
If HUND = 250 (1 sec rollover),
HUND = 0
Increment SECOND.
If SECOND = 60,
SECOND = 0.
Decrement time (MINUTE, HOUR)
If time = 0:0, set TFLAG.
If Override time 0:0, decrement override time.
(Update display.)
Turn all displays off
Increment DISPLY.
If DISPLY = 4, DISPLY = 3 (DISPLY counts 0–3).
If MODE = 0 (Powerfail),
If BLFLAG (Time to blink display),
If DISPLY = 0, convert minutes ones to 7-seg and
write to display reg.
If DISPLY = 1, convert minutes tens to 7-seg and
write to display reg.
If DISPLY = 2, convert hours ones to 7-seg and
write to display reg.
If DISPLY = 3, convert hours tens to 7-seg and
write to display reg.
If MODE = 1 (normal timekeeping),
If Override time = 0:0 (OVHOUR = OVMIN = 0),
If DISPLY = 0, convert minutes ones to 7-seg and
write to display reg.
If DISPLY = 1, convert minutes tens to 7-seg and
write to display reg.
If DISPLY = 2, convert hours ones to 7-seg and
write to display reg.

If DISPLY = 3, convert hours tens to 7-seg and
        write to display reg.
    Else (override time was >> 0:0),
        If DISPLY = 0, convert OVMIN ones to 7-seg and
            write to display reg.
        If DISPLY = 1, convert OVMIN tens to 7-seg and
            write to display reg.
        If DISPLY = 2, convert OVHOUR ones to 7-seg and
            write to display reg.
        If DISPLY = 3, convert OVHOUR tens to 7-seg and
            write to display reg.
If MODE = 2 (time set),
    If DISPLY = 0, convert PRMIN ones to 7-seg and
        write to display reg.
    If DISPLY = 1, convert PRMIN tens to 7-seg and
        write to display reg.
    If DISPLY = 2, convert PRHOUR ones to 7-seg and
        write to display reg.
    If DISPLY = 3, convert PRHOUR tens to 7-seg and
        write to display reg.
Now update the discrete status LEDs.
If MODE = 0 or 1 (powerfail or normal timekeeping),
    Turn off SET LED.
    If ONOFF (timing ON time)
    OR if OVTIME >> 0 and VOFLAG set (ON override),
        If not MTFLAG (water level OK),
            Turn on ON/OFF LED.
        Else (MTFLAG set, water level low),
            If BLFLAG, turn on ON/OFF LED (Makes LED blink).
    If not ONOFF
    OR if not VOFLAG, turn off ON/OFF LED.
If MODE = 2 (Time Set),
    Turn on SET LED.
    If SEMODE = 1,
        If not MTFLAG (Water level OK),
            Turn on ON/OFF LED.
        Else (MTFLAG set, water level low),
        If BLFLAG, Turn on ON/OFF LED (Makes LED blink).
    If SEMODE = 0, turn off ON/OFF LED.
If PB switches are all off,
            Set DBCOUN = 0
            Set ONFLAG = 0
            Set OFFLAG = 0
            Set FCFLAG = 0
            SEFLAG = 0.

Else (a PB is pressed),
  If DBCOUN << 4,
    Increment DBCOUN.
    If DBCOUN = 4 (debounce done),
      Set flag corresponding to pressed switch.
If water low switch active,
  If MTCOUN << 255,
    Increment MTCOUN
    If MTCOUN = 255 (filter timeout), set MTFLAG.
  Else (water low switch inactive) clear MTFLAG.
  End of 250 Hz timer interrupt code.

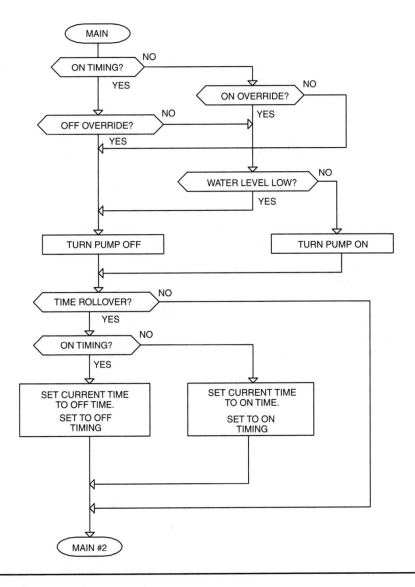

Figures A.1 (above), A.2 (p. 262), A.3 (p. 263)
Pool Timer Polling Loop Flowcharts.

Figures A.4 (p. 264) and A.5 (p. 265)
Pool Timer Schematic.

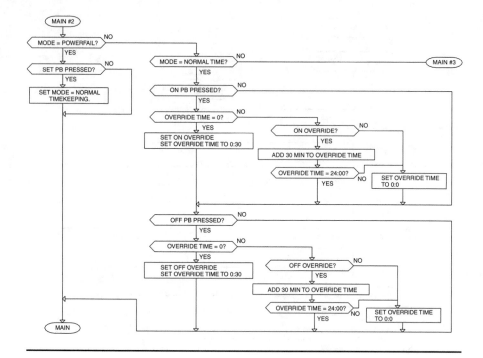

Figure A.2

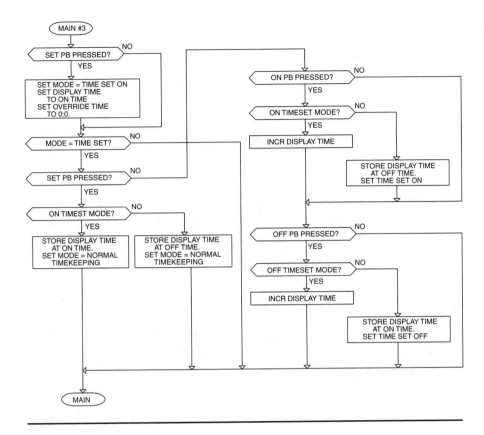

Figure A.3

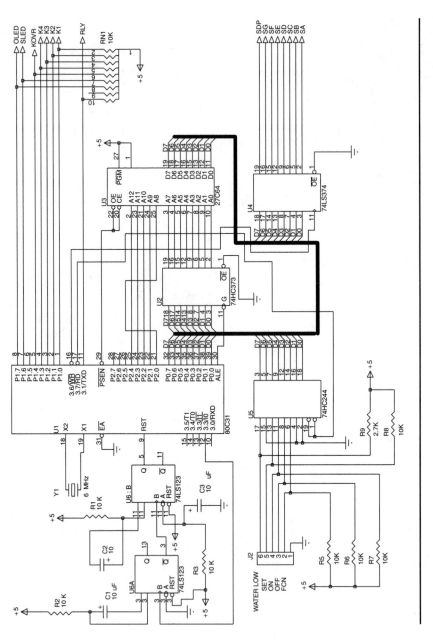

Figure A.4

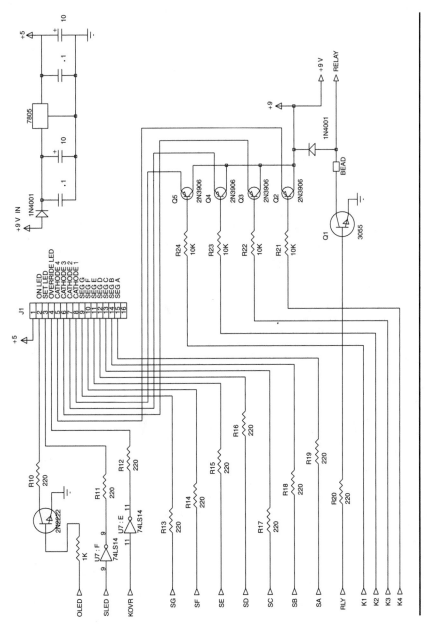

Figure A.5

Appendix B:
Number Systems

This book assumes knowledge of certain basic concepts. For the second edition I added this appendix and the following two, where some of these concepts are briefly reviewed. This limited space cannot serve as a thorough treatment of these topics, but the essentials will be covered.

Number Bases

Before looking at computer numbering systems, we quickly review the decimal system. If we have a four-digit number like 1234 we can write it this way:

$$(4 \times 1) + (3 \times 10) + (2 \times 100) + (1 \times 1000)$$

As we move from right to left in a decimal number, each digit is the next power of 10. The least significant digit, in the ones position, is 4. This is multiplied by 10^0, ($10^0 = 1$). The digit 3 is in the tens position, multiplied by 10^1. The 2 is in the hundreds position, multiplied by 10^2. Finally, the 1 is in the thousands position, 10^3. As you can see, the exponent of 10 starts at 0 in the rightmost digit, and increases by one for every digit you move to the left. Hence, 10 is the *base* of the decimal system.

The digits in any decimal number can range from 0 to 9. Since the decimal system is base 10, there are ten possible digits, including 0. This is necessary because any number system needs a unique character for every possible value in a single digit. When working with different number bases, it is common to use a subscript to indicate what the number base is. So 1234 in decimal would be written 1234_{10}.

Microprocessors use digital, or binary, logic, where everything is 1 or 0. Since there are two digits in a binary system, the base is 2. A binary number looks like this:

$$10011010010$$

Each position, or digit, in a binary number is called a *bit* (*B*inary dig*IT*). Just like the decimal system, each binary digit is an increasing power as you move from right to left. Only in this case, each position represents an increasing power of *2* instead of 10. The rightmost digit is in the ones position (2^0), the next digit is the twos position (2^1), the next digit is in the fours position (2^2), and so on. We can rewrite the binary number to show what value each bit corresponds to:

Original Number	1	0	0	1	1	0	1	0	0	1	0
Power of 2	2^{10}	2^9	2^8	2^7	2^6	2^5	2^4	2^3	2^2	2^1	2^0
Value of bit	1024	512	256	128	64	32	16	8	4	2	1

So, our example binary number can be calculated as:

$$(0 \times 1) + (1 \times 2) + (0 \times 4) + (0 \times 8) + (1 \times 16) + (0 \times 32) + (1 \times 64) + (1 \times 128)$$
$$+ (0 \times 256) + (0 \times 512) + (1 \times 1024)$$

or

$$2 + 16 + 64 + 128 + 1024$$

which is 1234_{10}.

Computers typically work with binary values that are 8, 16, 32, or 64 bits in length. Here, 8 bits can represent a value from 0 to 255 $(1 + 2 + 4 + 8 + 16 + 32 + 64 + 128)$. And, 16 bits can represent a value from 0 to 65,535; 32 bits can represent values up to 4.29×10^9; and 64 bits can go up to 1.84×10^{19}.

Obviously, writing numbers in binary is inconvenient for the human programmers who use the computer, so computer values typically are written in *hexadecimal* format. In hexadecimal, usually abbreviated *hex*, the binary word is separated into 4-bit groups. Grouped, the example value looks like this:

$$10011010010 = 0100\ 1101\ 0010$$

The two numbers are the same, but spaces are added to the second number to separate it into 4-bit groups, the way commas sometimes are added to decimal numbers. Note that an extra 0 was appended to the left of the leftmost group to make it a full 4 bits wide. Now, remember what the value of each binary bit position was and you can calculate the number this way:

$$10011010010 = 0100\ 1101\ 0010 =$$
$$(0 \times 1) + (1 \times 2) + (0 \times 4) + (0 \times 8) \quad = 2 = 2 \times 1$$
$$+ (1 \times 16) + (0 \times 32) + (1 \times 64) + (1 \times 128) \quad = 208 = 13 \times 16$$
$$+ (0 \times 256) + (0 \times 512) + (1 \times 1024) = 1024 \quad = 1024 = 4 \times 256$$

This is the same as what we had before, except that we are finding the sum of each 4-bit group (2, 82, and 1024) and adding those sums to get the total. What about the factors at the end of each line? Why is it important that $82 = 13 \times 16$? This is why:

$$2 = 2 \times 1 = 2 \times 16^0$$
$$82 = 13 \times 16 = 13 \times 16^1$$
$$4 \times 256 = 4 \times 16^2$$

So the word can be written $(4 \times 16^2) + (13 \times 16^1) + (2 \times 16^0)$.

As you can see, when we break the binary word into 4-bit groups, each group is an increasing power of 16 as you move from right to left. Each 4-bit group represents a digit of a base-16 number. The 4-bit groups make it base 16 because each 4-bit group can represent a maximum value of 15 $(1 + 2 + 4 + 8)$. Including 0, this makes 16 possible values for each digit. After 15, the number carries over to the next digit, just like decimal digits do when you reach 9.

Now take another look at those 4-bit groups. For the moment, we will treat each 4-bit group as an individual binary number and calculate them this way:

$$0100 = 4 = (0 \times 1) + (0 \times 2) + (1 \times 4) + (0 \times 8)$$
$$1101 = 13 = (1 \times 1) + (0 \times 2) + (1 \times 4) + (1 \times 8)$$
$$0010 = 2 = (0 \times 1) + (1 \times 2) + (0 \times 4) + (0 \times 8)$$

Just to clarify what we are doing, we rewrite the original grouped binary number with the corresponding values:

Binary:	0100	1101	0010
Decimal:	4	13	2

Notice that these are the same values we multiplied by the powers of 16 when we first broke the number into 4-bit groups:

Binary:	0100	1101	0010
Decimal:	4×256	13×16	2×1

So our original binary number can be written as a three-digit, base-16 number (4, 13, 2).

The only problem with this is how to write the numbers. We need a single character to represent each digit, even those greater than 9. Otherwise, we cannot tell the difference between the digit value 13 and the two digits 1 and 3. Since the decimal system cannot represent digits greater than 9, the alphabet characters A–F are used for the extended digits, like this:

Decimal	Binary	Hex
0	0000	0
1	0001	1
2	0010	2
3	0011	3
4	0100	4
5	0101	5
6	0110	6
7	0111	7
8	1000	8
9	1001	9
10	1010	A
11	1011	B
12	1100	C
13	1101	D
14	1110	E
15	1111	F

We now can write the number in three different bases:

$$1234_{10} = 10011010010_2 = 4D2_{16}$$

Since many text editors (especially those from the early days of computers) cannot handle subscripts, the numbers often are written without subscripts. Instead, a *b* suffix is used for binary and an *h* for hex, like this:

$$1234 = 10011010010b = 4D2h$$

Sometimes a lowercase *d* suffix is used to indicate decimal numbers, but if this method is used the hex numbers always must use uppercase digits (ABCDEF). Otherwise you cannot tell if the *d* indicates a decimal number or the hex digit D.

It is important to remember that microprocessors do not operate with hex numbers, they operate in binary. Hexadecimal is just a convenient representation for people to use when working with binary numbers.

In the early days of computers, *octal* often was used. This was just another representation, where the binary numbers were divided into groups of 3 bits. Each group could range from 0 to 7, and the digits went up by powers of 8 (1, 8, 64, etc.). 1234_{10} = 2322_8. Octal rarely is used now.

Why use 4-bit groups? Why not create a number system that uses 5-bit groups, where the values ranged from 0 to 31? You could, and it would let you represent numbers up to 1023 with two digits. But you would need 32 unique characters for the digits, and you have to remember the value of each of them. More important, microprocessor data and (usually) address buses come in increments of 8 bits, so the hexadecimal system is more practical for real systems.

Converting Numbers Between Bases

We often need to convert numbers between hex, decimal, and binary. The simplest way, of course, is to use a calculator that can convert between bases. However, it is important to understand the methods.

Hex to Binary

Hex to binary conversions are easy. Start with $1DE6_{16}$. To convert this to binary, just write out the binary values that correspond to each hex digit:

Hex	1	D	E	6
Binary:	0001	1101	1110	0110

If you want, take out the spaces to get 0001110111100110.

Now you can see why hex is easier to use. This is just a 16-bit number. Imagine working with 64-bit numbers using binary.

Binary to Hex

Separate the binary number into 4-bit groups, starting with the rightmost digit. If the rightmost group does not have 4 bits, append zeros to the *left* to make 4:

1110111100110 becomes   1 1101 1110 0110
Append zeros on the left: 0001 1101 1110 0110

Then convert each 4-bit group to a hex digit:

Binary:	0001	1101	1110	0110
Hex	1	D	E	6

Hex to Decimal

Factor each hex digit by the corresponding power of 16 and sum the results:

$$1DE6 = (1 \times 16^3) + (D \times 16^2) + (E \times 16^1) + (1 \times 16^0)$$
$$= (1 \times 4096) + (13 \times 256) + (14 \times 16) + (6 \times 1)$$
$$= 4096 + 3328 + 224 + 6$$
$$= 7654_{10}$$

Decimal to Hex

Divide the number by 16, write down the remainder. Divide the integer portion of the previous result by 16, write down the remainder. Continue this process until the division results in 0. Write the remainders in reverse order for the hex equivalent of the decimal number:

$$7654/16 = 478 \text{ with a remainder of } 6$$
$$478/16 = 29 \text{ with a remainder of } 14 \ (14_{10} = E_{16})$$
$$29/16 = 1 \text{ with a remainder of } 13 \ (13_{10} = D_{16})$$
$$1/16 = 0 \text{ with a remainder of } 1$$

Write the remainders, in hex, in reverse order: 1DE6.

Math with Binary and Hex Numbers

Binary numbers (and their hex representations) can be added and subtracted just like decimal numbers. Where most people get into difficulty is in the carry process. When you add two decimal digits, say, 9 and 7, you get 16. However, the process of doing this addition involves a carry:

$9 + 7 = 6$ with a carry of 1. The 1 carries into the next, or tens, digit.

Similarly, binary numbers have carry properties:

$$0 + 0 = 0, \text{ no carry}$$
$$0 + 1 = 1, \text{ no carry}$$
$$1 + 1 = 0, \text{ with a carry to the next binary position.}$$

So, if we add 9 and 7 in binary, it looks like this:

$$9 = 1001_2$$
$$+7 = 0111_2$$

We start by adding the least significant digits:

$$1 + 1 = 0, \text{ with a carry into the next digit.}$$

Adding the next pair of digits (in the twos position):

$$0 + 1 + 1 \text{ (the carry from the last add)} = 0 \text{ with a carry}$$
$$\text{Fours digit: } 0 + 1 + 1 \text{ (carry)} = 0 \text{ with a carry}$$
$$\text{Eights digit: } 1 + 0 + 1 \text{ (carry)} = 0 \text{ with a carry}$$

This is illustrated next:

$$
\begin{array}{r}
\text{Carry: } 1111 \\
1001 \\
+0111 \\
\hline
10000 = 16.
\end{array}
$$

We can add hex numbers the same way:

$$9 + 7 = 0 \text{ with a carry of } 1, \text{ or } 10_{16} \text{ or } 16_{10}$$

In hex, a carry occurs when the sum of two digits exceeds F_{16}. The following are a couple of examples:

$$
\begin{array}{r}
\text{Carry: } 100 \\
0205_{16} = 517_{10} \\
+0E07_{16} = 3591_{10} \\
\hline
100C_{16} = 4108_{10}
\end{array}
$$

$$
\begin{array}{r}
\text{Carry: } 111 \\
2678_{16} = 9848_{10} \\
+AAAA_{16} = 43690_{10} \\
\hline
D122_{16} = 53538_{10}
\end{array}
$$

Negative Numbers and Computer Representation of Numbers

In the examples so far, we worked with 4-bit values and added digits as needed when a result grew beyond that. In a computer, numbers are represented as a multiple of hex digits, usually 2, 4, or 8 digits. The number of digits depends on the word size of the computer (most microprocessors can concatenate words to make bigger values but that is unimportant for this discussion). An 8-bit machine will use two hex digits, a 16-bit machine will use four digits, and a 32-bit machine will use eight digits. So the value $2A_{16}$ would be represented this way:

$$\text{8 bit: } 2A$$
$$\text{16 bit: } 002A$$
$$\text{32 bit: } 0000002A$$

This may seem like an insignificant point, since all three numbers are the same. The only difference between them is the number of leading zeros in front of the significant digits. However, the word width is important when dealing with negative numbers.

Binary and hex numbers can be subtracted the same way as decimal numbers. However, in computer hardware, subtraction is difficult to accomplish. Negative numbers are difficult to store, since there is no place for a minus sign. In a computer, subtraction usually is performed by *adding* a *negative* number. A negative number is indicated by having the most significant bit as 1. This is why the word width is important. On an 8-bit machine, 80_{16} is not the same as 0080_{16} on a 16-bit machine. On the 8-bit machine, 80_{16} represents a negative value.

Negative numbers can be represented in one's complement or in two's complement form. A *one's complement* number is formed by complementing all the bits in the number:

$$0010\ 0111\ 1011\ 0101 = 27B5_{16}$$
$$\text{one's complement} = 1101\ 1000\ 0100\ 1010 = D84A_{16}$$

Note that the most significant bit is set, indicating that this is a negative number. You can do math with one's complement numbers, like this:

$$\text{Hex: } 3010_{16} - 27B5_{16} = 3010_{16} + D84A_{16} = 1085A_{16} = 085A_{16} = 2138$$
$$\text{Decimal: } 12304 - 10165 = 2139$$

Two notes about this: When we did the addition, the result was 1085A, but we threw away the leading 1, leaving a result of 085A. This is because we are working with a 16-bit (four hex digits) value. In a real computer, any carry beyond four digits is lost. The second thing to note is that the actual result we got, 2138_{10}, is 1 less than the right answer, 2139.

What happens if we do a subtraction and the result is negative? We use the same example, but subtract the larger number from the smaller one:

$$27B5_{16} - 3010_{16} = 27B5_{16} + CFEF_{16} = F7A4_{16} = -85B_{16} = -2139_{10}$$

Note that the result of the addition, F7A4, had the most significant bit set, so we know it was negative. Taking the one's complement of the result gives us the answer, 2139. The rules for one's complement math are as follows:

A number to be subtracted is made negative and added.

To make a number negative, invert each bit in the number.

Add the two numbers.

Throw away any carry beyond the number of digits you are using.

If the most significant bit (MSB) of the result is set, the result is negative.

If the MSB of the result is not set, add 1 to the (positive) result.

Since we had to add 1 to the original result to get the right answer, why not make that part of the number we subtract? That is exactly what two's complement is. To make a *two's complement* number, invert each bit in the number and add 1.

Two's complement: 27B5, inverted = D84A. Add 1, result = D84B.

Now we do that subtraction again, in two's complement:

$$3010_{16} - 27B5_{16} = 3010_{16} + D84B_{16} = 1085B_{16} = 085B_{16} = 2139_{10}$$

Try the version that gives a negative result:

$$27B5_{16} - 3010_{16} = 27B5_{16} + CFF0_{16} = F7A5_{16} = -85A_{16}$$
(Inverting 3010_{16} produces $CFEF_{16}$, and adding one gives $CFF0_{16}$)

Notice that the answer, F7A5 is correct, since it is the two's complement of 2139_{10}. So with two's complement, we need not add 1 to positive results. The result always is right, no matter what.

What happens if we add two negative numbers? Try this example:

$$-1010_{16} - 2010_{16} = EFF0_{16} + DFF0_{16} = CFE0_{16}$$

CFE0 is the two's complement of 12,320, which is the right answer. So the rules for two's complement math are

To make either number negative, invert all the bits and then add 1.

Add the two numbers.

If the MSB is 0, the result is positive and correct.

If the MSB is 1, the result is negative and correct in two's complement form.

Overflow

As already mentioned, math on a computer is limited to the word within use. If you try to add 60,000 and 60,000, you get 120,000. On a 16 bit computer, you'll get this:

$$60,000_{10} = EA60_{10}; EA60_{16} + EA60_{16} = 1D4C0_{16} = D4C0_{16}$$

What happened to the 1 in the most significant position? It was dropped because this is a 16-bit (four digit) system and we cannot represent numbers larger than 65,535. In fact, this addition turned the two positive numbers into a negative number. A computer that thinks it is working in two's complement will interpret this result (D4C0) as a negative number, specifically $-11,072_{10}$. This is called *overflow.*

A 16-bit word can represent values from 0 to 65,535 ($FFFF_{16}$). However, if the most significant bit is used as a sign bit, then the same 16-bit word can represent values from only $-32,768$ (8000_{16}) to $+32,767$ ($7FFF_{16}$). There still are 65,536 values, but half of them are negative. Note that the most negative value is not $FFFF_{16}$. The most negative value is 8000_{16}. $FFFF_{16}$ is negative 1, and it is what you get if you start with 0000 and subtract 1 (try it).

When you do math on a computer, the hardware does not necessarily know that you are using two's complement. When the MSB is treated as a sign bit, the number is said to be a *signed* number. When the MSB is part of the number, you cannot have negative values and the number is called an *unsigned* number. So if you want to add

30,000 and 30,000, you can treat the result ($EA60_{16}$) as an unsigned, positive result ($60,000_{10}$) or as a signed, negative number (-5536_{10} in two's complement).

Of course, with a wider word (32 or 64 bits), the range of values, both positive and negative, is much greater.

Number Suffixes

One final word about the hexadecimal number system involves the abbreviation K (for kilo). When you see the suffix K attached to a number in electronics or finance it implies a multiplier of 1000. A 1 K resistor, for example, is 1000 ohms; 100 K dollars is $100,000. However, in the computer world, K means "multiply by 1024." So a 16-bit-wide word can have 65,536 possible values, or 64K (65,536/1024 = 64).

A similar rule applies to the term *meg*, or million. A 1 meg resistor is 1,000,000 ohms. In computer lingo, a meg is 1024 × 1024, or 1,048,576.

Floating Point

A limitation on any integer number scheme, regardless of the number of bits, is the difficulty in representing fractional numbers such as 2.54 or 3.3. When we looked at decimal numbers at the beginning of this appendix, we saw that they increase in powers of ten as you move from right to left across the digits. As you move to the *right* of the decimal point, decimal numbers increase in *negative* powers of 10:

10^2	10^1	10^0	...	10^{-1}	10^{-2}	10^{-3}	10^{-4}
or 100	10	1	...	.1	.01	.001	.0001

Binary numbers work the same way:

	2^2	2^1	2^0	...	2^{-1}	2^{-2}	2^{-3}	2^{-4}
Decimal	4	2	1	...	.5	.25	.125	.0625

And hex numbers as well:

	16^2	16^1	16^0	...	16^{-1}	16^{-2}	16^{-3}
Decimal	256	16	1	...	.0625	.0039	.000244

So we can write a decimal number, such as 2.54, in binary and hex:

$$2.54 = 010.100010100011_2 = 2.8A3_{16}$$

Note that, in binary and hex, the number is a repeating value. Just like one third is a repeating decimal in base 10 (but not in base 3), some fractional numbers cannot be exactly converted between bases.

We could represent fractional binary numbers in a computer by defining a 16-bit number as ranging from 0 to 4095 instead of 0 to 65,535. The 4096 values can be represented by the upper 12 bits of the 16-bit word. This leaves the lower 4 bits available to represent fractional values. For instance, the hex value 1002 would be interpreted as 100.2, or $100_{16} + 2 \times 16^{-1}$, or 256.125 in decimal.

Such an arrangement makes calculations fairly easy and keeps everything in an integer format. However, the resolution of the fractional part of the number is limited, and a trade-off must be made between the accuracy of the fractional part and the maximum size of the number. The more bits allocated to the fractional part, the smaller the maximum number can be. The fewer bits allocated to the fractional part, the less precision we have to represent numbers with.

A better means of representing fractional values would emulate the decimal system with which we already are familiar. With four decimal digits, you can write .0001, 10, or 1000. All these numbers use four digits, but the decimal point can move, or *float*, to represent different values. This is the concept behind floating point numbers. A floating point number typically is represented in a computer in this format (16 bits shown here):

$$s\ eeeeee\ fffffffff$$

where s is a sign bit (0 = positive, 1 = negative), eeeeee is the exponent (6 bits), and fffffffff is the mantissa (9 bits), always positive.

We can represent all the eeeeee bits, collectively, as E. We can represent all the fffffffff bits, collectively, as F. Then the value of the number is given by

$$-1^S \times F \times 2^E$$

Now we can represent any number within the range of the exponent (−31 to +31). Note that, to represent fractional values, we must be able to use a negative exponent. The exponent is biased so that a value of 0 (in this example) represents an exponent of −31. An exponent of all ones (111111) represents +31. In effect, you take the binary value of the exponent field and subtract 31 from it to get the actual exponent. If we were using a 7-bit exponent, we could represent values from −63 to +63, and we would have a bias of −63. This allows representation of negative exponents without needing to resort to two's complement math.

For our example, a 0 in the exponent field represents an exponent of −31, a value of 25 represents an exponent of −6 (25 − 31), and a value of 44 represents an exponent of 13. Remember, these are exponents of 2, not exponents of 10.

So, using 9 bits for F, we can represent our 2.54_{10} in fractional binary as

$$10.100010100011_2 = 10.1000101 \text{ (truncate at 9 bits)}.$$

When working in bases other than decimal, the decimal point is called a *radix point*. We shift the value to the right so we always have a number of the form 1.xxxx and add an exponent:

$$1.01000100 \times 2^{-1}$$

We always arrange binary numbers so that they take the form 1.xxxx. Since this is the case, we can throw away the 1 and gain another bit of precision to the right of the radix point:

$$.010001010$$

This is read as $(1 + 2^{-2} + 2^{-6})2^1$ or 2.53_{10}. The leading 1 is implied.

The obvious question is, How can we guarantee that the number always can be represented by 1.xxxxx so we can drop that leading 1? If you think of scientific notation in decimal, you can represent any decimal number as d.ddddd × 10^{yy}, where d stands for any decimal digit and yy is an exponent (positive or negative) of 10. Even very small numbers can be represented this way, by using a large negative exponent. The same rules apply to binary. The only difference is that, in decimal, we have no way to know what the digit to the left of the decimal point is. In binary, we know it has to be 1. What if the entire number is 0? We get to that later.

Now we can create a 16-bit floating point number from our example value:

0	100000	010001010
sign	exponent	mantissa, leading 1 implied

The general steps for converting a decimal number in the form xxx.yyy to floating point format are these:

- Convert xxx (digits to the left of the decimal point) to binary (call it aaaa). Convert yyy (digits to right of decimal point) to fractional binary, call it bbb. Write as a fractional binary number:

aaa.bbbb

- Shift the number to the right, keeping track of the exponent, until there is a single 1 to the left of the radix point:

a.aabbb (6-bit example shown, works for any size number)
exponent = 2 (because we shifted two positions right)

- Drop the leading 1 and calculate the exponent using the bias of the exponent field. If the number is positive, make the sign bit 0. Otherwise, the sign bit is 1.

The IEEE has developed a standard for representation of floating point numbers. The IEEE format defines single- and double-precision values. The IEEE single-precision format uses a sign bit, 8 exponent bits, and 23 mantissa bits, for a total of 32 bits. The double-precision standard uses 64 bits: a sign bit, 11 exponent bits, and 52 fractional bits. The single-precision exponent can range from −127 to +127, and the double precision exponent can range from −1023 to +1023.

Finally, what do we do if we have 0? Zero cannot be represented by 1.xxxx. The IEEE standard defines 0 as being represented when the exponent and mantissa both are 0. The sign bit can be either.

The IEEE standard also reserves the maximum exponent value (FF for single precision, 7FF for double precision) to indicate an overflow condition; numbers that are either too small or too large to be represented.

Appendix C:
Digital Logic Review

This appendix reviews digital logic concepts. The review will not be comprehensive but will address those portions of the topic that are needed in the book. The concepts presented here refer to basic digital logic gates and functions, even though those functions usually are implemented in some type of programmable or configurable logic in modern designs.

The basic concept behind digital logic is ones and zeros. A digital logic signal is either one or zero, high or low, on or off. The high/low, on/off state may be defined in different ways. For TTL logic, high is anything over about 2.4 volts, while a low is anything below 0.8 volts. In between is an undefined region where the signal is never supposed to be.

For CMOS logic operating at 5V, the high/low cutoff is about 2.5 volts—anything higher is considered "high" anything lower is considered "low." An RS-232 signal, like the ones that come from the COM ports on a PC, swing both positive and negative. The high state is anything above +3V, and the low state is anything below −3V. A current-loop interface, like the MIDI signals that connect music synthesizers, defines *high* as the absence of current flow in a pair of wires, and *low* as the presence of current flow.

Differential logic is unique in that the high/low state can be defined with only two signals. If one is at a higher voltage than the other, the resulting state is "high." If the two are reversed, the result is "low." If both are the same, the signal state is undefined.

Sometimes digital signals are described as true or active and false or inactive. In this case, the true/active and false/inactive states may be defined as either high or low. When working with microprocessors it is quite common to find signals that are true or active in the low state.

A signal that is high usually is capable of driving (sourcing) some current into whatever is connected to it. A signal that is low usually is drawing, or sinking, current from the connected device. Typical digital logic circuits cannot sink current when in the high state or source current in the low state. In some cases, such as CMOS logic, the impedance of the receivers is very high and the amount of current is insignificant except when the signal is changing states. However, the sourcing-while-high and sinking-while-low restriction still applies to the driving device, even if the receiving devices are neither using or providing current. If two outputs are connected together and one is low while the other is high, the output is indeterminate. In real logic, the low output usually wins, but the voltage is not guaranteed to be a valid logic state. Whichever output wins, both will have considerable current flowing through them, and one or both often is damaged if the condition persists. Connecting outputs in this way is not considered a valid design practice. When this condition occurs it is called *output contention* or *bus contention*. The term *output contention* usually refers to a single signal and *bus contention* refers to a group of signals, such as a microprocessor data bus.

Some digital devices can sink current in the low state but do not source current in the high state. These usually are the same as their current-sourcing siblings but without the transistor in the output stage that sources current. If the logic is a bipolar family, such as TTL, these outputs are referred to as *open-collector*. If the logic family is CMOS, these outputs are called *open drain*. Open-collector-/-drain outputs are designed to be tied together. If one output goes low while the other is high, no damage will occur since the high output does not source current. Open-collector-/-drain signals normally are pulled high with a resistor so that the signal will be in a valid high state when none of the outputs is driving it low.

Basic Logic Functions

Simple Gates

Figure C.1 shows some simple logic gates. The simplest digital logic gate is the inverter. An inverter inverts whatever is applied to the input. If a 1 is applied to the input, a 0 appears at the output and vice-versa. Note the "bubble" at the output of the inverter. This indicates that the signal is inverted. If no bubble were present, this symbol would indicate buffer, not an inverter, and the output would follow the input.

The AND gate is another logic function. It has two or more inputs. If both inputs are high, the output is high (A *and* B). If *either* input is low, the output is low. Although the figure shows a two-input gate, the AND gate can have many inputs. However many inputs it has, the logic works the same way; all inputs must be high for the output to be high. If any input is low, the output is low.

The OR gate also has two inputs, but the output of an OR gate is high if either input is high (A *or* B). The output is low only if both inputs are low. Like the AND gate, the OR gate can have many inputs. As long as one input is high, the output will be high.

Variations on the AND and OR gates are NAND and NOR gates. The NAND gate is an AND gate but with the output inverted. If any input is low, the output is high; and if all inputs are high the output is low. The NOR gate is an OR gate with the output inverted. If any input is high, the output is low; and if all inputs are low, the output is high. Like AND and OR gates, NAND and NOR gates can have more than two inputs.

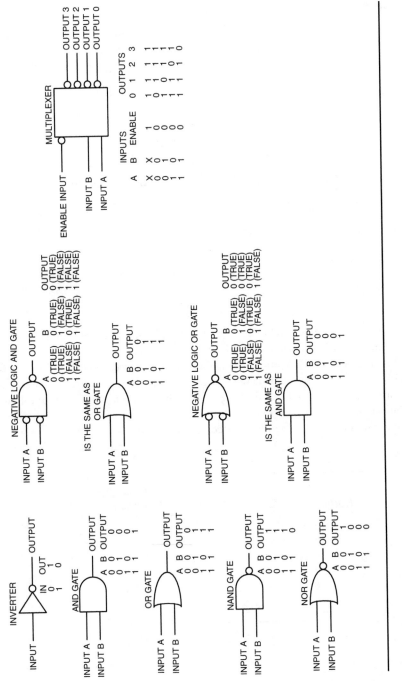

Figure C.1
Basic Logic Gates.

Don't Care

Sometimes in digital logic, the *don't care* state is a valuable designation. The don't care state, usually designated by X, indicates that the state of the signal does not matter— it will not affect the output. With the AND gate shown in the figure, input B is a don't care state as long as input A is low:

	Normal Logic Table			Logic Table Using Don't Care	
A	B	Output	A	B	Output
0	0	0	0	X	0
0	1	0	X	0	0
1	0	0	1	1	1
1	1	1			

X = don't care.

You can see that the logic table is the same for both cases. As long as A is low, the output is low, regardless of what state B is in. Similarly, as long as B is low, the output will be low, regardless of A. What this illustrates is an inhibit capability—if input A is a signal that constantly switches between high and low, we can control whether the signal appears at the output by controlling input B. While B is high, the output follows A. While B is low, the output is low.

A similar don't care table can be created for the OR gate:

	Normal Logic Table			Logic Table Using Don't Care	
A	B	Output	A	B	Output
0	0	0	0	0	0
0	1	1	X	1	1
1	0	1	1	X	1
1	1	1			

X = don't care.

In this case, holding B high forces the output high, and taking B low allows the output to follow A.

True/False Notation

As already mentioned, we can define the input and output as true and false instead of high and low. If we do this with the basic AND and OR gates, we get the following:

	AND Gate			OR Gate	
A	B	Output	A	B	Output
False	False	False	False	False	False
False	True	False	False	True	True
True	False	False	True	False	True
True	True	True	True	True	True

All we did here was call 0 false and 1 true. These tables are the same as the original logic tables for the AND and OR functions.

Negative Logic

Logic functions such as AND, NAND, and OR also can be used in an inverting configuration, where a true is 0 and false is 1. This typically is indicated with inversion bubbles at the input and output, as shown in Figure C.1. The logic of the invert-AND gate would be like this:

If A is low AND B is low, the output is low.

If either A OR B is high, the output is high.

Note that this is the same as the logic for the original OR gate. Negative logic reverses the function of the gates. A low/true AND function is implemented with an OR gate, and the low/true OR function is implemented with an AND gate.

Tristate

One more basic logic function needs to be described, *tristate*. In the tristated (sometimes called *high-impedance*) condition, the driver does not drive the signal—it neither sinks nor sources current. The voltage floats to some unknown level, or if the signal is pulled up with a resistor, the signal will go to a high state. A tristate output has three states: high, low, and tristated.

Tristate is essential to microprocessor designs. A typical microprocessor will have a common group of 8, 16, 32, or 64 signals for reading and writing data. When signals are grouped this way, they are referred to collectively as a *bus*. When the processor wants to write data, it drives the data bus with the data it wants to write, and all other devices connected to the bus are expected to tristate their drivers so there is no conflict. When the microprocessor wants to read data from the bus, it tristates its own signals and the device that it wants to read from is expected to drive the bus with the requested data. Tristate signals allow many digital outputs to be tied together, but the basic rule still applies—only one device at a time can drive the signal.

Tristate devices come in two flavors: unidirectional and bidirectional. A unidirectional device can send data in only one direction—to the output. The output can be either high, low, or tristated, but it is never an input. The outputs of a bidirectional device can also be tristated, but they double as inputs, allowing data back into the device. Again, when the outputs of a bidirectional device are tristated, they also act as inputs and can receive signals from another device that is driving the shared signal. Microprocessor data buses always are bidirectional, since they are used for both reading and writing. Most microprocessor peripheral ICs are bidirectional as well.

A common use of bidirectional ICs in microprocessor circuits is as bus buffers. A microprocessor data bus will be connected to one side of a bidirectional driver IC (called a *transceiver*). Call that side A. Some other device will be connected to the other side of the transceiver, side B. When the transceiver is off, it drives neither bus. When side A (connected to the microprocessor) is enabled, the signals on side B appear on the microprocessor data bus and the microprocessor can read them. When the side B outputs are enabled, data from the microprocessor bus is passed to side B. Transceivers typically are available in 8- or 16-bit widths to accommodate common microprocessor buses.

Multiplexers

In microprocessor designs, multiplexers are normally used for address decoding. As shown in figure C-1, a multiplexer has multiple outputs, but only one at a time is active. Multiplexers normally have low-true outputs like the example in the figure, and they usually have an enable line that makes all the outputs false. The multiplexer in the figure has four outputs, selected with two inputs (A & B). Multiplexers are also available with 8 outputs and 3 select inputs. Of course, if a multiplexer function is implemented in a PLD or other configurable logic device, it may have any number of outputs with any polarity (even mixed) and the enable may not be required.

Set/Reset Flip-Flop

These are *storage* devices. A flip-flop remembers its state. A typical flip-flop will have two inputs, *set* and *reset*, and an output, Q. When set goes low, Q goes high. Q then stays high regardless of which state the set input goes to. Q does not go low until the reset input goes low. Q then stays low until set goes low again. Flip-flops can be constructed with high/true or low/true inputs and inverted or noninverted outputs.

Figure C.2 shows the logic symbol and timing diagram for a set/reset flip-flop. As indicated, this type of flip-flop can be built using a pair of NAND gates. Only one output is shown in the figure, but the output of the other NAND gate also can be

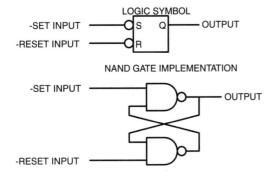

LOGIC SYMBOL

-SET INPUT ——— S Q ——— OUTPUT
-RESET INPUT ——— R

NAND GATE IMPLEMENTATION

-SET INPUT ——————— OUTPUT

-RESET INPUT ———————

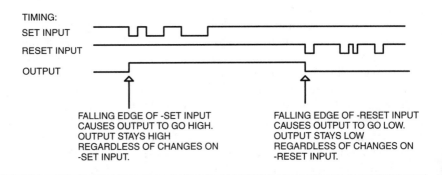

TIMING:
SET INPUT
RESET INPUT
OUTPUT

FALLING EDGE OF -SET INPUT
CAUSES OUTPUT TO GO HIGH.
OUTPUT STAYS HIGH
REGARDLESS OF CHANGES ON
-SET INPUT.

FALLING EDGE OF -RESET INPUT
CAUSES OUTPUT TO GO LOW.
OUTPUT STAYS LOW
REGARDLESS OF CHANGES ON
-RESET INPUT.

Figure C.2
Set/Reset Flip-Flop.

used and will be an inverting output. A pair of NOR gates wired the same way as figure C-2 also will function as a flip-flop, but the inputs will be high/true instead of low/true.

So what happens if both inputs to a NAND flip-flop go low at the same time? If you look at the logic, both NAND gates have one input low, so both outputs will go high. However, this condition is not latched, and when one input goes back high, the corresponding output will go back low. If both of the inputs go high at the same time, the final state of the output will be indeterminate. A similar result occurs in a NOR flip-flop if both inputs are taken high—both outputs go low.

Registers and Latches

Microprocessor circuits invariably require some kind of registered logic. This often is embedded in the peripheral ICs connected to the processor. However, often some form of registered logic must be included in a microprocessor circuit, to latch an address or function as a latched output port.

Appendix C   285

There are several types of registers and latches, but the types most commonly used in microprocessor circuits are D-type latches and D-type registers.

A D-type device passes the data input to the output. The output may be noninverted, in which case the output will follow the input, or it may inverted, where the output is the inverse of the input. In either case, the device exhibits storage, like a flip-flop. The output will "remember" what state it was in, even when the input goes away. So, if the input sent the output to a 1, then the output will stay a 1 even when the input changes state.

The control over what the output does is performed with a latch or clock input. A D-type registered device has a clock input and will transfer the input to the output (called *capturing*) on the rising edge of the clock. When the clock is in any other state (low, high, or falling), the output does not change state, regardless of input changes. It is possible to build a D-type device that captures on the *falling* edge of the clock, but these are not commonly used.

What happens if the input is changing while the clock is rising? This results in a *race condition*, and the output will be indeterminate. Actual devices have a minimum setup time, measured in nanoseconds, that the input must be stable before the clock changes to guarantee a valid output.

A latched device has a latch input (commonly called *G*) and it will pass the input to the output as long as the latch is high. This is called the *transparent* mode. Any changes on the inputs will be reflected at the output while the latch input is high. When the latch goes low, the input is captured. The output does not change as long as the latch remains low. D-type latches typically are used to capture the address on a multiplexed microprocessor data bus. Like the registered device, the latched device requires that the data be stable for some number of nanoseconds before the latch goes low. If this requirement is not met, the output will be indeterminate.

Figure C.3 shows the timing characteristics of the two types of latches.

Registers and latches commonly are packaged in 8- or 16-bit versions to match microprocessor data buses. When packaged this way, all the latches or registers in the package are clocked to a common clock or latch pin.

Latches and registers also are available with tristate outputs, where a common output enable pin enables all the outputs in the package. There even are devices that combine a transceiver and latch into a single package, making a bidirectional, latched (or registered) transceiver.

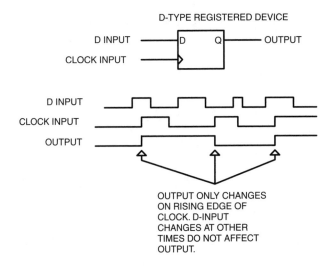

D-TYPE REGISTERED DEVICE

D INPUT ——— D   Q ——— OUTPUT
CLOCK INPUT ———

D INPUT
CLOCK INPUT
OUTPUT

OUTPUT ONLY CHANGES
ON RISING EDGE OF
CLOCK. D-INPUT
CHANGES AT OTHER
TIMES DO NOT AFFECT
OUTPUT.

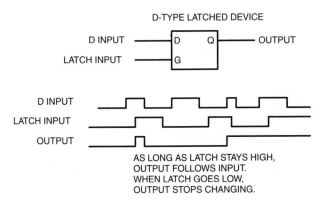

D-TYPE LATCHED DEVICE

D INPUT ——— D   Q ——— OUTPUT
LATCH INPUT ——— G

D INPUT
LATCH INPUT
OUTPUT

AS LONG AS LATCH STAYS HIGH,
OUTPUT FOLLOWS INPUT.
WHEN LATCH GOES LOW,
OUTPUT STOPS CHANGING.

Figure C.3
Registered Devices.

Appendix D:
Basic Microprocessor Concepts

A microprocessor is a compact computer. Early microprocessors were much simpler than the typical minicomputers and mainframes of the day, but many modern microprocessors are more complex and more powerful than computers of that era. Dozens of different microprocessors are available from many manufacturers, and they vary in speed, power, size, and capability. Regardless of the complexity, though, the basic architecture at the heart of all microprocessors is the same.

A Simple Microprocessor

The core of a microprocessor is the *arithmetic logic unit*, or ALU. The ALU takes in two values and produces a result. The result can be the sum of the two input values, the difference, a logical result (ANDing or ORing all the bits together), or some other operation. Which function is performed is determined by control inputs to the ALU. Figure D.1 shows a simple ALU that operates on two inputs, X and Y, producing a result. The inputs and the output can be any number of bits, 1, 4, 8, or 16.

This ALU can perform four functions: addition, negation, logical AND, and logical OR. Addition is a simple, binary, mathematical addition. Negation inverts all the bits in the input variable, making zeros into ones and ones into zeros. The AND function ANDs the bits of the two variables, making any given output bit 1 only if both corresponding input bits are ones and 0 otherwise. The OR function performs a bitwise OR, making the output bits 1 if either of the corresponding input bits is 1, 0 only if both inputs are 0. These operations are illustrated next with 4-bit values:

Variable A:	1001
Variable B:	1100
A ORed with B:	1101
A ANDed with B:	1000
A negated:	0110
B negated:	0011

Where do the numbers come from? Figure D.2 shows an expansion of the ALU concept, adding two banks of four registers each. Each of the eight registers has the

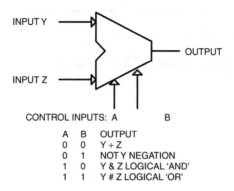

A	B	OUTPUT
0	0	Y + Z
0	1	NOT Y NEGATION
1	0	Y & Z LOGICAL 'AND'
1	1	Y # Z LOGICAL 'OR'

Figure D.1
Simple ALU.

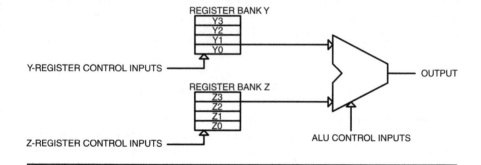

Figure D.2
ALU with Register Banks.

same number of bits as the inputs and output of the ALU. Now the ALU can get data from four registers on each side. It can add register Y0 to register Z3 or AND Y2 with Z1. Two control inputs to each bank of registers allow selection of any register in the bank. If we were building this with discrete logic, each register would consist of an 8-bit D-type register with tristate outputs. All the D0 bits would be connected together, all of the D1 bits connected together, and so forth. To read any of the registers, the output enable would be driven low to place the register contents on the ALU input (but only one register at a time in each bank). But, what do we do with the output of the ALU? And, how does data get put into the Y and Z register banks?

Figure D.3 shows an additional connection; the output of the ALU is connected to the inputs of the register banks. Now the ALU can add the contents of two registers and store the result back into one of the registers in either bank. Of course, to make this work, each register will need a clock. Figure D.3 shows a timing diagram of how

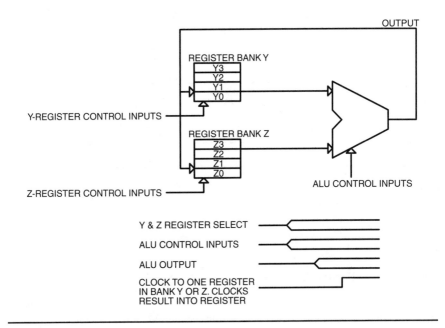

Figure D.3
ALU with Connection to Register Inputs.

such a system might work. As you can see, the register select inputs go to some value to select one register in the Y bank and one in the Z bank. The outputs of the selected registers are applied to the inputs of the ALU.

At the same time as the register select signals go active, the ALU control signals go active to select which ALU function will be performed. After some propagation delay through the ALU, the output reflects the result of the selected operation. Some time after that, a clock signal clocks the result into one of the registers in the bank. Only one register at a time gets a clock.

Control Store

So far, we have left out any discussion of where the control signals come from. Figure D.4 shows the addition of timing logic and a *control store*. The control store contains a sequential list of "instructions" that our simple computer operates on. In this simple system, the control store could have a bit assigned to each function. This would require two bits each for the ALU control bits and the register select for each bank. Three additional bits would be needed to select into which of the eight registers the result is to be clocked. The control store is like the Y and Z register banks, in that it contains data and the input determines which register contents will be applied to the output. The difference is that the control store cannot be written to, only read from.

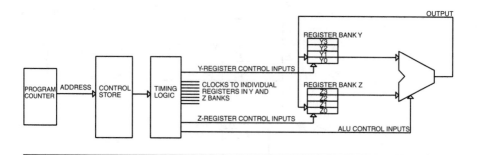

Figure D.4
Addition of a Control Store.

Addressable Memory

The control store is one type of addressable memory. An addressable memory has an input and an output. The input is a binary number, and the output is a different binary number.

You can think of addressable memory as being like a row of apartments. Somebody named Tom lives in apartment number 1. Frank lives in apartment number 2, and Zoe lives in apartment number 3. If you stand at the end of the hall and shout for whomever lives in apartment 1 to come out, Tom will step in to the hall. If you shout for the person in apartment 3 to come out, you will see Zoe.

Now suppose that the people in the apartments have numbers instead of names. Tom is 117, Frank is 145, and Zoe is 4567. Now if you shout for the person in apartment 1 to come out, number 117 will appear. Note that our hypothetical apartment complex can have only one person living in each apartment.

The apartment number in this simple example is equivalent to the address that is input to an addressable memory. Each location (apartment) in the memory has a number (person) stored there. When the address of a location is applied to the input of the memory, the number stored in that location appears at the output.

The numbers in the memory need not be unique. Just as you can have two Toms living in the same apartment complex, you can have multiple instances of the same number in different locations of an addressable memory.

One difference between apartment numbers and memory locations is zero-based addressing. Apartment and house numbers do not start with 0, but memory locations do. Remember that the input to an addressable memory is a binary number, and all zeros is as good a binary number as any other. In a microprocessor system all the addresses are used, including 0.

The address and output need not be the same number of bits. For example, an addressable memory may have a 10-bit address (1024 locations) and an 8-bit output (256 possible values).

Real addressable memories have other inputs in addition to the address, we look at those later. The concept of addressable memory is key to understanding how microprocessors work.

Timing Logic

The timing logic will not be examined in detail. It just makes sure that things happen at the right time, such as waiting until the ALU outputs are stable before clocking them into one of the registers.

Program Counter

The control store is driven by a *program counter*. This is just a binary counter that counts from 0 to however large the control store is. If the control store holds four instructions, the program counter needs to be 2 bits wide. If the control store holds 1024 instructions, the program counter has to be 10 bits wide. The program counter is incremented each time an instruction is executed, to select the next instruction in the control store.

Opcodes

Say that the control store is 9 bits wide and we define the bits like this:

Bits 0, 1: select ALU function (00 = addition, 01 = negation, 10 = AND, 11 = OR).

Bits 2, 3: Select Z-register (00 = Z0, 01 = Z1, 10 = Z2, 11 = Z3)

Bits 4, 5: Select Y-register (00 = Y0, 01 = Y1, 10 = Y2, 11 = Y3)

Bits 6, 7, 8: Select which register the result will be clocked into:

000 = Z0, 001 = Z1, 010 = Z2, 011 = Z3,
100 = Y0, 101 = Y1, 110 = Y2, 111 = Y3

The different combinations of bits that tell the machine what to do are called *opcodes*. Now, say we want to write a program that will execute the following two operations:

Add Y1 to Z2, putting the result in Y2

AND Y2 with Z3, putting the result in Y3

The control store would contain the following words, based on the preceding bit definitions:

Location	Control Store
0	110 01 10 00
1	111 10 11 10

The program counter starts at 0 (remember, zero-based addressing), and the output of the control store causes the first operation to be executed. Then the program counter increments to 1 and the second operation is executed.

Branching

Now we can write a program, up to the length of the control store, to do any operation that our simple machine is capable of. But, what happens when we get to the end of the program? To handle this, we can expand our control store from 9 to 20 bits, as shown in Figure D.5.

Of the added 11 bits, 10 wrap back around to the program counter as inputs, and 1 is a branch control bit. When the branch control bit is 0, the program counter increments, just as it did before. But when the branch control bit is 1, the program counter is loaded with the ten branch address bits from the control store. Now we can write a program that loops:

Location	Original Control Store	Branch Control Bit	Branch Address (10 bits)
0	110 01 10 00	0	X
1	111 10 11 10	1	0

After the first instruction is executed, the branch control bit is 0, so the program counter increments to location 1. But, after the second instruction, the branch control bit is 1, so the program counter does not increment. Instead, the branch address value, 0, is loaded into the program counter and the next instruction comes from that location. Our simple machine now will loop forever, executing these two operations.

Immediate Data

Note that when the machine is not branching, the bits in the control store that contain the branch address are not used and the value there does not matter. If we added more control bits, we could have an instruction that did not use the ALU, but instead clocked 8 bits of the branch address value into one of the registers. Now we have an *immediate* data instruction that can initialize the registers directly from the control store.

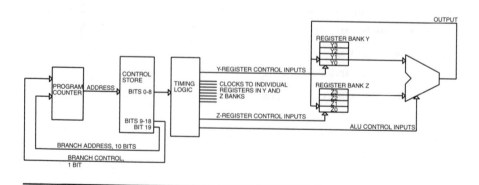

Figure D.5
ALU with Branching Capability.

Conditional Branching

We can expand this concept farther by adding more branch control bits. Two bits would let us have four branch options, such as not branch, branch always, branch if all the ALU outputs are 0, or branch if the result of an addition overflowed. Of course, we would need additional logic to detect the zero and overflow conditions.

Output

We are almost finished with this example, but there is one more step. We have a machine that can execute up to 1024 instructions (10-bit program counter), but what do we do with the results? Figure D.6 shows a final addition to the machine that adds a simple output scheme. This change adds a bank of four output registers. If the ALU has 8-bit inputs and outputs, then this bank of registers provides 8×4 or 32 bits of output. The outputs from the timing logic that control clocking into the Y and Z registers are expanded to add clocks to the control register. We can control the added outputs by making the 3-bit field that controls which Y or Z register gets the ALU output into a 4-bit field. The control store bit definition now looks like this:

Bits 0, 1: select ALU function (00 = addition, 01 = negation, 10 = AND, 11 = OR).

Bits 2, 3: Select Z-register (00 = Z0, 01 = Z1, 10 = Z2, 11 = Z3)

Bits 4, 5: Select Y-register (00 = Y0, 01 = Y1, 10 = Y2, 11 = Y3)

Bits 6, 7, 8, 9: Select which register the result will be clocked into:

0000 = Z0, 0001 = Z1, 0010 = Z2, 0011 = Z3,
0100 = Y0, 0101 = Y1, 0110 = Y2, 0111 = Y3
1000 = OR0, 1001 = OR1, 1010 = OR2, 1011 = OR3

Plus the ten branch address bits and one branch control bit.

In a similar way, we could expand the bit fields that select the Y or Z inputs to the ALU, so that we could enable one or more 8-bit tristate buffers instead of the internal registers. This would give the machine the ability to input information from the outside world.

Now we have a complete, although very simple, microprocessor. A real microprocessor works much the same way, but it includes the following improvements:

- Much more complex, capable ALU functionality. This typically includes more logic functions such as exclusive-OR, logical shifts, and other capabilities.
- A larger program counter, 16 to 64 bits wide. However, some microcontrollers with small internal PROMs may only have a 10- or 12-bit program counter, like our example.
- More complex branching conditions. These might include branching on overflow, branch on ALU carry, branch on some input bit being 1 or 0, and so on.
- More complex control store definitions. Our simple machine used a fixed control bit definition. For example, bits 2 and 3 always define which Z-register will be used. A real microprocessor might have instructions that do not use some registers. An

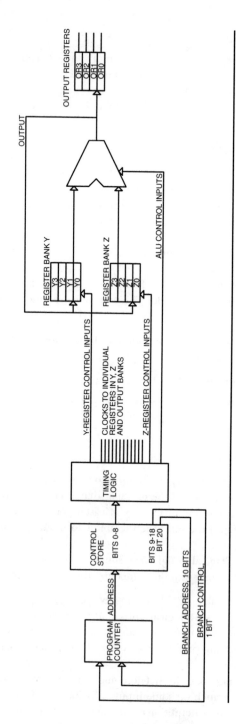

Figure D.6
Simple Microprocessor with Output Capability.

immediate instruction might load data directly from the control store to one of the registers. None of the bits that select which register drives the ALU is needed, nor are the ALU control bits needed. So, for those instructions, these bit definitions would change. A branch instruction might use the Z-register bits to determine what branch condition to test for (carry, no carry, zero, nonzero, etc.). We looked at a simple case of this, with the possibility of allowing the control store branch address field to double as a data value for nonbranching instructions. Making the control bits perform different functions for different instructions complicates the timing and control logic, but allows the control store word to be implemented with fewer bits.

- A microcontroller may have an internal program store, like our example, but many processors provide the program counter outputs on an external address bus so the control store can be outside of the microprocessor IC.
- In addition to internal registers, a real microprocessor typically has a means to produce an address that allows an external register bank, or memory, to be accessed. This address bus usually is shared with the program counter address bus. Our simple example could simulate this by defining an output register as an address register and another output register as a data register. The machine would write a value to the address register, then write the desired data to a data register. A data register clock also would have to be provided, so the external memory knows when a write has occurred. If the output registers are 8 bits wide, this would permit access to an external memory of 256 locations. A real microprocessor typically can access a much larger external memory and allows the address to be part of the instruction. This is called an *immediate* address, similar to the immediate data field we looked at earlier. In our simple machine, this probably would be the part of the control store bits that hold the branch address. Obviously, in this case, a branch instruction could not be an immediate instruction and vice-versa. It is up to the external memory device to decode the 8-bit address and determine to which specific register (or location) to write.
- A real microprocessor often can perform *indirect* operations, where the address of external memory or the external control store to be derived from an internal register. These registers often are incremented or decremented automatically as part of the instruction.
- The ability to branch to an address contained in a register. In our simple machine, this would require another path from one of the register banks back to the program counter inputs.
- The ability to link two registers together for some operations. Two 8-bit registers may be linked to make a 16-bit-wide memory address register. Typically, increment and decrement operations operate on the register as a single 16-bit value.

More complex microprocessors have other, more sophisticated features, but this covers the basic components that go into a modern microprocessor or microcontroller.

A More Complex Microprocessor

Figure D.7 shows a block diagram of another simple microprocessor that incorporates some of the preceding features. In this diagram, we have a microprocessor IC that contains an ALU, register bank, accumulator register, timing logic, instruction register, indirect address register, address mux, data mux, and program counter. Outside the microprocessor itself we have two devices: an external control store and an external memory.

The ALU is like the ALU in our earlier, simple machine. It performs arithmetic and logical functions on the values at the inputs. This is a 16-bit machine, with a 16-bit wide ALU.

The output of the ALU drives a register bank with four registers. Results of ALU operations can be clocked into any register of the bank, and any register in the bank can be used as one of the operands in ALU operations.

The other ALU operand always comes from the accumulator register. This is a single 16-bit-wide register. The accumulator-based model is common in many simple microprocessor designs. Typically, other registers are in the microprocessor, but the accumulator will be the only one that can be directly tested for zero or parity or some other logical condition. In many microprocessors, the accumulator is the only register on which some operations can be performed, such as increment and decrement. Other microprocessors allow nearly any operation to be performed on almost any register.

The timing logic gets information from the instruction register and controls the timing of the other blocks. This includes loading and reading the registers, incrementing and loading the program counter, and selecting the address mux source.

The instruction register receives information from the external control store and memory. Instructions are stored here as well as data.

The indirect address register was not in the simple processor we looked at earlier. The indirect address register is a register that can be loaded with the results of an ALU operation. The output of the indirect address register drives one of the address mux inputs.

The address mux is a device with three 16-bit input buses, one output bus, and control inputs. The address mux is controlled by the timing logic and can place the program counter contents, the indirect register contents, or the instruction register contents onto the 16-bit external address bus.

The data out register just captures the contents of the ALU result bus, to drive the external data bus for external write operations. This allows data to be written to the external memory.

Finally, the program counter is just like the program counter in the simple microprocessor, but it has the ability to be loaded from the ALU result bus.

This microprocessor has three external connections: a 16-bit address bus, a 16-bit data bus, and a control bus that consists of select signals for the external control store and the external memory. On a typical microprocessor, the individual address lines would be A0 through A15 and the data lines would be D0 through D15. Note that the address bus is output only, but the data bus can both send data from the microprocessor and receive data from the external devices.

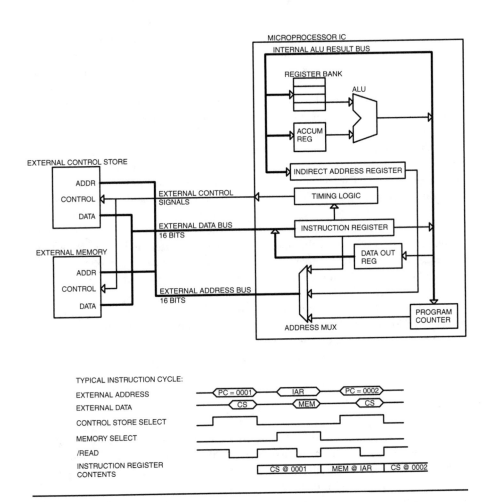

Figure D.7
A More Complex Microprocessor.

The external control store is just like the control store in the simple system, but it is outside of the microprocessor chip. The external memory is readable and writable. The control bus consists of two signals, /READ and /WRITE (the slash, /, indicates that the signals are true when low).

Figure D.7 also shows a timing diagram of how this microprocessor accesses the two external devices. Say that the program counter (PC on the diagram) starts out at location 0001. The timing logic, knowing that an instruction needs to be fetched, sets the address mux to place the contents of the program counter on the external address bus. After some setup time, the /READ signal is driven low, also controlled by the timing logic.

When the address was placed on the bus, the control store recognized that it was being selected. For simplicity, say that the control store recognizes any address from 0000 to 7FFFh, and the memory recognizes any address from 8000h to FFFFh. For now, ignore *how* the two devices know to which address to respond. When the /READ signal goes low, the control store places the contents of location 0001 onto the external data bus and the data are clocked into the instruction register at the end of the bus cycle. The program counter also is incremented, so it now contains 0002.

Say that this instruction opcode tells the processor to get the 16-bit word of memory pointed to by the indirect address register (IAR) and load it into one of the registers in the register bank. The timing logic decodes the 16-bit value in the instruction register (the one loaded from address 0001 in the control store) and initiates this operation.

First, the address mux is configured to pass the contents of the IAR register onto the external address bus. Say this address is A105h. The external memory recognizes this address, so when /READ goes low, the memory places the contents of A105h onto the data bus. At the end of the bus cycle, the data is clocked into the instruction register.

Now the next location in the program counter is passed to the bus, and the contents of that location in the control store are clocked into the instruction register. At the same time, the contents of the instruction register, which was loaded from memory on the previous bus cycle, are clocked into whatever destination register it is supposed to go to.

This simple example is very similar to a real microprocessor. A few things are worth noting:

- First, the external address bus is 16 bits, so it can access 65,536 (64 K) locations. In our example, the control store uses half and the memory uses half, so each has 32,768 locations. However, there is no reason the control store could not be 48 K in size and the memory 16 K or vice-versa.
- The control store and the memory are identical, except that the memory can be written to as well as read from. This means we could use one device to do both functions, as long as we do not overwrite the area where the instructions are stored. Instead of a 32 K control store and a 32 K memory, we could use a 64 K memory with the instructions stored in the lower half and data stored in the upper half. In fact, this is what many systems do, including the PC you probably have on your desk or at home. The PC has a small amount of memory that can only be read (like our control store) and a huge amount of memory that can be both read and written. The read-only memory is used to start everything, and then everything the computer needs to run is loaded from the disk drive and stored in the read/write memory.
- The external data bus can be either the contents of the IAR register, the contents of the program counter, or the contents of the instruction register. This implies that we can perform only one operation at a time (get an instruction, get data, write data, etc.). It is possible to build a microprocessor with multiple address buses that can perform more than one kind of bus cycle at a time to different storage devices.

- Although we did not walk through an example, the program counter can be loaded from the ALU output or from the instruction register. So we could add two numbers and make the sum the *address* of the next instruction we execute. Or, we could have an instruction that is followed by a data byte, and the data byte is the new starting point for the program counter. This gives the microprocessor branching capability like the simpler machine we looked at earlier. We even could have an instruction that uses the contents of the IAR to get a data value that is the address of the next instruction. This would be an indirect branch instruction.
- The contents of the instruction register can be placed on the ALU bus and loaded into one of the registers or the program counter. This implies the ability to tristate the outputs of the ALU. In a real microprocessor, the tristate function probably would be performed by a multiplexer like the address mux, because tristating buses inside the chip requires more logic. But the effect is the same.
- The second bus cycle used the contents of the IAR to address the external memory device. If the IAR had held a value between 0000h and 7FFFh, the control store would have been selected instead, and we would have read the data from there. So we could dedicate a portion of the control store to a table of data. This data could be almost anything that is constant, such as a degrees-to-sine conversion table or a table of atmospheric pressure versus altitude.

The timing logic is a complex digital system. It controls the following functions:

- Decoding the opcode in the instruction register.
- Selection of which source will be passed to the external address bus (based on the opcode).
- Timing the external /READ and /WRITE signals and determining (based on the opcode) whether the external bus cycle will be a read or write cycle.
- Remembering whether the contents of the instruction register are an opcode or data that need to be put someplace.
- Determining (based on the opcode) which register in the register bank will provide an input to the ALU and which register will be clocked with the data on the ALU result bus at the end of the instruction (accumulator, register bank, IAR, or PC).
- Determing (based on the opcode) which ALU operation will be performed.
- Incrementing and loading the program counter.

Finally, we look at the issue of how the two external devices knew to what addresses to respond. Real memories have read and (for writable memories) write inputs. They also have a signal that selects the memory. This signal can be generated by logic that decodes the address bus. In a simple system like this, the memories could just use the highest address bit (A15) since there are only two devices. The control store would respond when A15 is low and the read/write memory would respond when A15 is high. In a more complex system, additional gating logic decodes the address bus and generates select signals for all the external devices.

Addressing Modes

Here, we consolidate the various methods used to address memory in a microprocessor system, including those we already have looked at. Figure D.8 illustrates these addressing modes. For this section, we assume we have a simple microprocessor like the one in Figure D.7, with a 16-bit data path and 64 K memory space. We look at the effects of various addressing and branching modes on the processor program counter (PC in the diagram) and on two internal registers, R0 and IAR.

In the example shown in the figure, immediate data follows the instruction opcode in memory. Instructions that need no additional data are followed by another opcode. It is up to the microprocessor timing logic, which decodes the opcode, to remember that the following byte is data and not another opcode. For these examples, we do not worry about what the specific opcode values are, just what the opcodes do.

Direct Addressing

In direct addressing, the instruction contains the information that will be used. In the example, the instruction opcode is followed by a data value that gets loaded into the IAR. In this example, the opcode (at location 0000) says, "load the immediate data value (following the opcode) into register IAR." The data value following the opcode (0010) is loaded into the IAR.

Indirect Addressing

Indirect addressing uses a register to point to the data. Continuing with our example, the processor executes the instruction at location 0002. This is an indirect instruction that says, "Put the value addressed (pointed to) by IAR into register R0." Since IAR contains 0010 and location 0010 contains 12AB, we end up with the value 12AB in register R0.

Direct Branching

Direct branching, like direct addressing, includes the destination address (new PC value) as part of the instruction. Our example system continues, executing the instruction at location 0003. This is a direct branch instruction that says, "Start executing at the location pointed to by the data value following the opcode." Since this data value is 0008, the processor loads the PC with 0008 and continues on.

Indirect Branching

Again, like indirect addressing, indirect branching takes the destination address from a register. In our example, the processor executes the instruction at location 0008, which loads the IAR with a new value (0015). The instruction at location 000A says, "Start executing at the location addressed by IAR." Since IAR contains 0015, the next instruction is fetched from there.

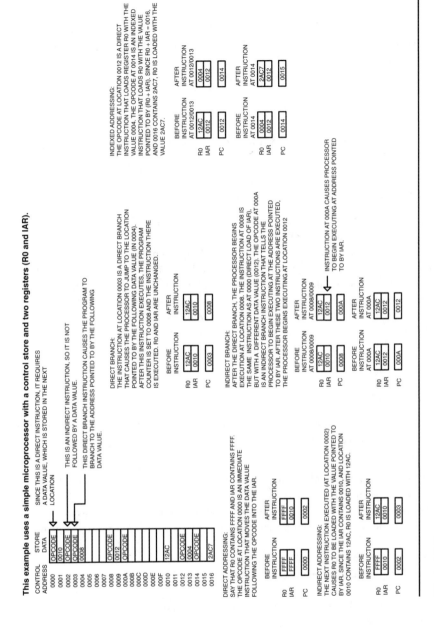

Figure D.8
Addressing Modes.

Indexed Addressing

Indexed addressing uses two values to access a location. In the example, register R0 is loaded with 0004 by the instruction at address 0012/0013. IAR already contains 0012 from the indirect branch instruction just executed. The instruction at 0014 is an indexed instruction that says, "Load R0 with the value addressed by [IAR + R0]." Since IAR + R0 = 0012 + 0004 = 0016, the value from 0016 (2AC7) is loaded into R0. Note that we loaded the value into one of the registers used to calculate the address; we could have loaded it into another register in the processor. Some microprocessors support an indexed-direct instruction, where one of the two index parameters is immediate data in the instruction.

Indexed Branching

Although not shown in the example, indexed branching works the same way as indexed addressing, where a pair of registers are added to generate the destination address. A special case of indexed-direct branching commonly is used in microprocessors, where a direct data value is used as an index from the program counter. The direct data value usually is a signed 8- or 16-bit number, allowing branching of ±127 (8-bit) or ±32 K (16-bit) locations. For example, if the program counter is 12BC, a branch instruction that contains an 8-bit value of 06 would cause a branch to 12C2 (12BC + 06). On some microprocessors, this is the only kind of conditional branch available and only unconditional branches have the ability to reach the entire address range of the CPU.

The Real World

Real microprocessors range from fairly simple devices, like this example system, to much more complex devices. Further enhancements that you might see on a real processor include:

- More registers. Some might have special functions, such as a stack pointer (stack pointers are discussed elsewhere in the book).
- Independent bus interface and execution units (the Intel ×86 family has this). This permits the bus to fetch a new instruction while an old one is executing, improving overall performance.
- Internal peripheral devices such as timers.
- Interrupt capability, where an external event can temporarily redirect program execution.
- Capability for another processor to control the bus, allowing multiple processors to share a single bus.

Code Formats

Getting instructions into the microprocessor means storing them in the control store memory in some way. The code (that is, the ones and zeros that get loaded into memory for the microprocessor to execute) is called *machine code.* Of course, writing programs in machine code would be very tedious. Every branch address would have to be calculated by hand, and if you needed to insert an instruction between two existing instructions, you have to recalculate all the addresses.

The next level up from machine code is *assembly.* Assembly code replaces the machine code with simple statements that are translated directly into machine code by assembler software. There is one assembly statement per microprocessor instruction. The assembler allows branches to be defined with *labels* (names), and the assembler calculates branch addresses. Assembler statements usually are abbreviations of the instruction functions.

A machine instruction that moves data between two registers, R1 and R2, might use an assembler statement like this:

```
MOV R1, R2    (MOVe R1 to R2)
```

A statement that moves an immediate value of 23 into register R1 might look like this:

```
MVI R1, 23    (MoVe Immediate value 23 to R1)
```

A branch instruction might look like this:

```
JMP label     (JuMP to address of label)
```

To insert a new line of code, you have to edit the *source file,* which contains the assembly statements. Then the assembler is run and new machine code is produced, which then is programmed into the control store. Assemblers always allow you to insert comments into the code to explain what you are doing. These may be preceded by a semicolon (;), double slash (//), or other characters.

Every microprocessor has a unique assembly language, although many manufacturers use a common language across a family of processors. The following is an example of assembler code and the corresponding machine code for an Atmel AVR-series microcontroller:

Machine Code	Assembler Code	Comments
94f8	cli	; CLI disables interrupts
ec0c	ldi accum,$CC	; Put CC (hex) into accum
bb05	out portc, accum	; Output accum to port C
ef0f	ser accum	; Set all the bits in the
		; accum register to ones.
bb04	out ddrc, accum	; Set port C to outputs
98de	cbi porta,6	; Clear port A bit 6

Machine Code	Assembler Code	Comments
0000	nop	; do nothing (delay)
c002	rjmp clk_tach_on	; Jump to a label called ; clk_tach_on

Finally, high-level languages provide a simpler, easier means of programming micro-processors. A high-level language, such as C, permits the programmer to write instructions that look like this:

```
x = y + z;  // Add two numbers
```

The *compiler* translates the instructions into machine code. Unlike assembly, there is not one high-level-language statement per machine instruction. One high-level line of code may generate dozens of machine instructions. The preceding example might produce machine code instructions that do this:

```
move memory location 'y' into Register 1
move memory location 'z' into Register 2
Add Register 1 to Register 2, leaving the result in Register 1
Store the contents of Register 1 in memory location 'x'
```

This simple C statement produced four lines of machine code. Using a high-level language, the software engineer need not worry about the specifics of the machine language or assembly language for the microprocessor. High-level languages permit better portability of the code across different microprocessors.

With this overview, you should be ready to tackle the material in the book.

Appendix E:
Embedded Websites

The following is a list of websites for manufacturers and organizations that produce embedded products. Of course, this list is not comprehensive and any listing of web addresses is subject to change at a moment's notice.

Organizations and Literature

CompactPCI systems: http://www.compactpci-systems.com

Embedded Systems: www.embedded.com

Embedded Technology: www.embeddedtechnology.com

PC/104 consortium: www.pc104.org

PC/104 supplier page: www.pc104.com

PCI industrial computer's manufacturer's group: www.picmg.com

VME bus international trade association: www.vita.com

Manufacturers

AMD: www.amd.com

Atmel: www.atmel.com

Dallas Semiconductor: www.dalsemi.com

Fujitsu: www.fujitsu.com

Hitachi: www.hitachi.com

Intel: www.intel.com

Microchip: www.microchip.com

Mitsubishi: www.mitsubishichips.com

Motorola semiconductors: www.mot-sps.com

SGS-Thompson: www.st.com

Sharp microelectronics: www.sharpmeg.com

Texas Instruments: www.ti.com

Toshiba: www.toshiba.com

Waferscale Integration: www.waferscale.com

Zilog: www.zilog.com

Software, Operating Systems, and Emulators

2500 AD software: www.2500ad.com

Accelerated Technology: www.atinucleus.com

American Arium: www.arium.com

Annasoft: www.annasoft.com

Applied Microsystems: www.amc.com

ByteCraft: www.bytecraft.com

CAD-UL: www.cadul.com

CMX: www.cmx.com

Embedded System Products: www.esphou.com

Embedded Systems Software: www.etcbin.com

Green Hills Software: www.ghs.com

Hi-tech: www.htsoft.com

Hitex: www.hitex.com

Huntsville Microsystems: www.hmi.com

IAR systems: www.iar.com

Kadak: www.kadak.com

Keil Software: www.keil.com

Mentor Graphics: www.mentorgraphics.com

Microsoft: www.microsoft.com

Nohau: www.nohau.com

QNX: www.qnx.com

SDS: www.sdsi.com

SMX: www.smxinfo.com

Softaid: www.softaid.com

Synapticad: www.syncad.com

Glossary

A/D: Analog-to-digital converter. An integrated circuit (IC) or subsystem that translates a voltage to a digital word.

Assembler: A language that directly describes machine instructions such as move data to a register, jump to an address, add two registers, and so on. Each microprocessor has a unique machine language, and therefore a unique assembler language.

Cache: A secondary memory used to reduce the bottleneck of memory access to a fast CPU. Data are moved from main memory into a faster cache memory and fetched from there. When the CPU needs data that is not in the cache, it must be fetched from the main memory.

CAN (controller area network): A multinode network using a single twisted-pair cable and capable of operating at speeds from 10 kbps to 1 Mbps. CAN originally was developed for the automotive industry.

CISC (complex instruction set computer): A computer that includes relatively complex instructions in the instruction set. CISC is a relative term. The instruction set of a CISC microcontroller may be much simpler and less flexible than that of a high-performance RISC CPU. *See* RISC.

Context switch: The context of a CPU usually refers to all the internal registers, including the stack pointer and instruction pointer. A context switch is the process of changing or restoring the CPU context to execute a different section of code (such as an interrupt service routine) and usually includes saving the current context.

CPU (central processing unit): Technically the computing core of a microprocessor; the term is commonly used to refer to the microprocessor itself.

Cross compiler (cross assembler): A compiler or assembler that runs on one computer but generates object code for another family of computers. An assembler that runs on a PC and generates code for a microcontroller is an example of cross assembly.

D/A (digital-to-analog converter): An integrated circuit or subsystem that translates a digital word to a voltage.

Daisy-chained interrupts: An interrupt prioritizing scheme where the priority of each peripheral is determined by its position in the chain. Lower-priority devices may acknowledge an interrupt only when no higher-priority devices are requesting an interrupt.

Debugger: *See* Monitor.

Device driver: Software that provides an interface between the operating system and actual hardware, such as video display boards or printers.

DMA (direct memory access): A mechanism whereby a microprocessor temporarily gives up its external bus to another processor (or other controller) and permits the other processor to directly access memory. Some microprocessors have built-in DMA controllers.

DRAM (dynamic RAM): RAM that stores information as charge on a capacitor. It must be periodically refreshed to renew the charge and retain data.

DSP (digital signal processor): A microprocessor optimized for processing signals such as sound, video, or radio frequency. A DSP typically includes hardware such as single-cycle multiply hardware, barrel shifters, and other features that are designed to speed signal processing.

Edge-sensitive interrupt: An interrupt that is recognized on a rising or falling edge.

EMC (electromagnetic compatibility): A general term for the measure of a device or system to operate in an environment with EMI. Usually used in relation to EMC testing or EMC standards.

EMI (electromagnetic interference): A general term for interference caused by ESD, radiated emissions, and magnetic interference.

EPROM (erasable programmable read-only-memory): A PROM that can be erased using ultraviolet light.

ESD (electrostatic discharge): Static electricity that is discharged to, inside, or around equipment.

Firmware: Software in machine-readable form, embedded in a ROM, PROM, EPROM, flash memory, or other nonvolatile storage.

Flash memory: A PROM that can be electrically erased and reprogrammed.

Harvard architecture: A microprocessor architecture where the code (instructions) is in a separate memory area from the data. A given memory address typically references different physical memory locations for code than for data.

HLL (high-level language): Any computer language that permits code to be developed above assembler. C, Pascal, and BASIC are high-level languages.

ICE (in-circuit emulator): A device designed to plug into a circuit and replace the target processor. A typical ICE permits the code to be run, breakpoints to be set, and the registers and memory of the system to be examined.

Interrupt controller: An integrated circuit or internal part of a microprocessor that prioritizes interrupts and provides a vector to the processor.

ISA (industry-standard architecture): The expansion bus and connectors used on the original IBM AT computer and still available on most x86-based PC motherboards.

ISR (interrupt service routine): Code executed when an interrupt occurs; it handles interrupt-specific functions.

Latency (interrupt): The time from when an interrupt occurs to when it is serviced.

Level-sensitive interrupt: An interrupt that is recognized while in the active state.

Machine language: The binary ones and zeros that the microprocessor reads from memory and executes. *See* Assembler.

Microcontroller: A microprocessor with internal RAM and I/O ports, and sometimes including ROM.

Microprocessor: An integrated circuit containing, at minimum, a central processing unit and a means to access external memory. Microprocessors also may include internal memory, I/O ports, or peripherals.

Modified Harvard architecture: A variation on the Harvard architecture where there is limited ability to obtain data from the code space. Many single-chip microcontrollers use the modified Harvard architecture.

Monitor: A program that executes on the target system and allows the engineer to examine memory and I/O, set breakpoints, and download code. It often supports other features as well. The term *debugger* is nearly synonymous with *monitor* and usually denotes a more sophisticated tool with advanced features.

ms (millisecond): One thousandth of a second.

Native development: Development of microprocessor code on the same family of CPUs as the code will be run on. Development of code on a PC to be run on a PC is native mode development.

Nested interrupts: Where interrupts are structured so that a lower-priority ISR can be interrupted by a higher-priority ISR.

NMI (nonmaskable interrupt): An interrupt input, available on many processors, that cannot be masked off. If the interrupt occurs, the processor always will service it.

ns (nanosecond): One billionth of a second.

NVRAM: A package housing a static RAM integrated circuit and a battery. The battery powers the RAM so that it will retain its contents when external power is off.

Object code: Code for a target system. It may be in binary or in some ASCII-hex representation of the data, such as Intel or Motorola hex formats.

OTP EPROM: One-time programmable EPROM. An EPROM without the erasure window. The OTP EPROM acts like a one-time programmable PROM but has an EPROM structure internally.

Overflow: A condition that occurs when the result of a mathematical operation cannot be represented by the number of bits available.

Passive backplane: A bus board that consists of only connectors, the interconnecting traces, and sometimes signal terminators. The CPU in a passive backplane system plugs into the backplane.

PC-104 bus: A bus architecture using pass-through pin/socket connectors. Electrically similar to the ISA bus.

Pipeline: A method of increasing processor throughput by prefetching instructions and storing them for the CPU to execute. Pipelining takes advantage of the time that the CPU spends executing instructions to buffer one or more additional instructions.

PLD (programmable logic device): A programmable integrated circuit used to implement logic functions.

Preemptive scheduling: A scheduling technique where each task is given control until it finishes or is superceded by a higher-priority task.

PROM (programmable read-only memory): A ROM that can be programmed, either by a PROM programmer or by the target system. Once programmed, acts as a read-only memory (ROM).

Protected mode: A memory-management mode available on some x86-family processors that provides hardware memory protection and other features.

Race condition: Any condition where two signals or events that happen simultaneously cause timing errors. The timings for hardware race conditions normally are measured in nanoseconds or microseconds. For software events, the timing can be any window within which the events appear simultaneous to the code.

RAM (random access memory): Memory that is both readable and writable and where any location may be accessed at any time. Memory locations in RAM do not need to be accessed in any specific sequence.

Real mode: A memory management mode on x86-family processors that segments memory into a maximum of 1 MB, with 64 K segments, and no hardware protection against invalid accesses.

Reentrancy: The ability of a section of code to be reentered without first finishing. Reentrancy requires that variables used in the code to be stored on a stack or with some other mechanism that prevents them from being overwritten when the code is reentered. Reentrancy is typically needed if a routine can be interrupted and then called by the interrupt service routine.

RISC (reduced instruction set computer): A computer that executes a simple, limited instruction set. The idea is that a simpler instruction set can be executed very fast, making up for the limited functionality with extreme speed. RISC is a relative term; a RISC microcontroller may be hundreds of times slower than a CISC computer. *See* CISC.

ROM (read-only memory): A memory device that can be read from but not written to.

RTOS (real-time operating system): Firmware that provides task scheduling, memory allocation, and other services for a real-time application.

Sequential (round-robin) scheduling: A scheduling technique where tasks are given control one at a time, in sequence, and each runs until finished.

Single step: A means, in either software or hardware, to cause a program to execute one instruction and then stop. Single stepping may be at the machine level, where one CPU instruction is executed, or at the level of a HLL, where one HLL statement (possibly many CPU instructions) is executed.

Skew: The condition that occurs when grouped signals (such as a microprocessor data bus) do not all change at the same time. This term also applies to differences in the logic paths inside a device, such as an address decoder. Even if the external signals change at the same time, differences in the internal delays may cause the same effect as if the external signals changed at different times. Skew usually is measured in nanoseconds.

Software: Computer instructions. This may refer to the source code or the actual machine-readable data.

SRAM (static RAM): RAM that is implemented as an array of flip-flops. Information is retained until overwritten or until power is removed.

STD bus: A bus architecture using a five-pin edge connector. Originally intended for 8-bit, 64 K processors, the STD bus has been expanded to include 16-bit processors and expanded addressing. STD-32 supports 32-bit processors and addressing.

Target: The system or microprocessor that an emulator is designed to install to or replace when debugging.

TCB (task control block): A memory area where an operating system stores information about tasks under its control.

Time slicing: A scheduling technique where a central scheduler switches tasks at regular intervals, giving each task in sequence a specified number of time slices to execute before going to the next task.

UART (universal asynchronous receiver/transmitter): An IC or circuit that provides an asynchronous serial interface.

Vector (interrupt): A number or instruction that is translated into an address, which then is executed to service an interrupt.

VME bus: A bus architecture based on one to three 96-pin DIN connectors. Originally designed around the Motorola 68000 processor timing.

von Neumann architecture: A microprocessor architecture where the code (instructions) can share the same memory space as the data. Most microprocessors intended for multichip designs use the von Neumann architecture.

WDT (watchdog timer): A timing circuit that resets or otherwise notifies a microprocessor if it is not triggered at periodic intervals.

Index

Page numbers followed by "t" denote tables; those followed by "f" denote figures.

Index

ROM, 161
Engineering specifications
    definition of, 4
    description of, 1
    function of, 4, 196
    for multiprocessor systems, 195–196
EPROM. *See also* Flash memory
    access time calculations, 41–43
    benefits of, 13
    components of, 36
    costs of, versus ROM, 14
    data hold time, 43
    description of, 13, 36, 38
    electrically erasable. *See* EEPROM
    erasing process, 38–40
    inputs for, 41
    output enable time, 43
    schematic representation of, 37f
Erasable programmable read-only
    memory. *See* EPROM
ESD
    definition of, 80
    protection methods, 81–82
Event-driven scheduling, 202–204
Exception
    definition of, 248
    handling of, 248

FIFO buffers, 176
Filters, for electrostatic discharge
    protection, 81
Firmware specifications, 1
Flash memory. *See also* EPROM
    access time calculations, 41–43
    advantages of, 38
    block-organized, 39
    device manufacturer identification
      by, 39
    erasing of, 38
    in-circuit programming of, 38, 76–
      77
    mechanism of operation, 39
    programming of, 38–40
    properties of, 38
    SRAM and, 231
    wait states and, 238
Flip-flop
    registers with, 170f, 171
    set/reset, 284–285

Floating-point calculations, 103, 275–
    277
Flowcharts
    description of, 95–96
    for pool pump timer system
      example, 261–263
Functional requirements, 1

Gating logic, 65f
Grounding, for electrostatic discharge
    protection, 81
Ground loops, 82

Hardware
    memory management, 246–248
    partitioning determinations, 21–23
    requirements estimations, 19–21
    specifications, 1, 19–21, 25
Harvard architecture, 15–16, 16f
Hex numbers, 271–272
High-level language, 101–104, 306

I^2C bus
    buffering of, 182
    characteristics of, 70–71
    development of, 71
    for interprocessor communication
      in multiprocessor systems, 181–
      183
    Microwire and, comparisons
      between, 72–73
    schematic representation of, 69f, 70–
      71
    speed of, 71
ICs
    combination, 231
    description of, 57
    functions, 57–60
    interface, 57
    peripheral. *See* Peripheral ICs
    SDRAM, 236
    timer, 57
Idle loop. *See* Polling loop(s)
In-circuit programming
    description of, 14–15, 76
    of flash memory, 38, 76–77
    schematic representation of, 77f
Instruction set, evaluation of, 12
Integrated circuits. *See* ICs

Intel
  80186, 63
  80188
    description of, 59
    interfaces, 63
  i960 VH processor, 238, 243
  timing for, 30
Interfaces
  description of, 7
  differential, 82
  8-bit, 63–66
  electrostatic discharge protection, 81
  I²C bus, 71
  ICs, 57
  JTAG, 246
  microprocessor selection and, 7
  Microwire, 71–73
  16-bit, 63–66
  32-bit, 241
Interleaving, 232–233, 234f
Interrupt(s)
  acknowledge, 135
  actions secondary to, 113
  bus cycles, 118
  daisy-chained, 118–119, 124–125
  debugging effects, 136
  definition of, 113
  description of, 9
  edge-sensitive, 113, 121–123
  estimating requirements for, 9
  external, 117
  externally vectored, 123–125
  function of, 9
  hardware for implementing, 116–117
  internal, 117
  latencies, 11
  latency, 132f
  level-sensitive, 113, 120–121
  low-priority, 134–137
  microprocessor selection and, 9
  multiple reads and, 133–134
  nested, 115–116, 125, 127–128, 145
  nonmasking, 119–120
  overusage of, 9
  prioritizing of, 115–116
  protection against, 98–99
  race condition and, 28, 131
  in real-time operating system, 209–210

  reasons for using, 136
  shared memory and, 129–130
  software for, 125
  stuck, 128–129
  terminal, 125
  timer, 117, 123, 124f, 132–133, 145
  when to use, 136–137
Interrupt controllers
  description of, 57, 113
  vector response to, 114
Interrupt service routine
  actions secondary to, 125
  data transfer to or from, 128
  description of, 113–114, 117
  mechanism of operation, 125–126
  in real-time operating system, 209–210
  timer resetting and, 131–132
Interrupt vectors
  address, 116
  description of, 114
  generation of, 115f, 116–117, 124
  source of, 119
I/O (input/output)
  control, 218
  digital, 53f
  LSI, 56, 58t
  microprocessor selection criteria and, 6–7
  pins, 6–7, 27
  ports, 6, 57, 108
  schematic representation of, 53f
  simple, 53f, 53–54
  strobes, 53–54
ISR. *See* Interrupt service routine

JTAG interface, 246

Kernel, 200, 213

Language. *See* Development language; High-level language
Latches
  D-type, 286
  for extended data hold time, 62–63
  for I/O, 56
  packaging of, 286
Latch input, 286
Level-sensitive interrupts

Source synchronization, of DMA
    controllers, 78
Specifications
  engineering
    definition of, 4
    description of, 1
    function of, 4, 196
    for multiprocessor systems, 195–
      196
  hardware, 1, 19–21, 25
  software
    description of, 110
    detailed types of, 20–21
    estimating of, 20
    example of, 110–112
    reasons for creating, 110
    summary overview of, 25
Speed
  cache memory for improving, 239
  of I²C bus, 71
  of microprocessor
    estimating of, 11–13
    pitfalls regarding, 12–13
SRAM
  characteristics of, 44
  DRAM and, comparisons between,
    47–50
  flash memory and, 231
  nonvolatile, 44
  write cycle timing, 45f
Stack
  definition of, 105
  function of, 105–106
  hardwired, 127
  microprocessor, 204–205
  registers saved on, 126
State diagram
  definition of, 93
  for pool timer system, 93, 94f
State machine(s)
  description of, 99–100
  incremental, 100
  multiple, 100
STD bus, 225–226
Strobes
  read, 47, 53
  write, 47
Superloop. See Polling loop(s)
Supervisor, 247

Switch closure, 108
Synchronization
  clock, for bus, 241–244
  destination, 78
  of distributed processor systems, 24
  of multiprocessor systems, 190
  source, 78
Synchronous bus, 34
Synchronous transfer, 35
System definition, 3–4

Task control block, 204
Tasks, in real-time operating system.
    See also Multitasking
  communication between, 205–207
  scheduling of, 207
  tracking of, 204–205
Test specifications, 1–2
Timer
  ICs, 57
  interrupts caused by, 117, 123, 124f,
    132–133, 145
  for pool pump timer system
    example, 261, 264–265
  in real-time operating system, 209
  watchdog
    description of, 75
    electrostatic discharge protection
      secondary to, 82
    functions of, 75
    mechanism of operation, 75, 76f
    sophisticated types of, 75–76
Time slicing
  definition of, 200–201
  sequential scheduling and, 201
Timing
  access
    for EPROM, 41–43
    for RAM, 44–47
  acknowledge, 189–190
  calculations, 52–53
  cumulative errors in, 131–133
  DMA, 73, 74f
  DRAM, 48, 50f
  interrupt effects, 127
  Microwire, 69f
  schematic representation of, 42f
  SDRAM, 237
  of software, 144–145

Timing logic
    description of, 293
    functions of, 301
Trace data, for debugging
    circular trace buffers, 145–147
    read from ROM, 143–144
    software timing and, 144–145
Transceiver, 284
Tristate, 283–284
True/false notation, 283

UART, 57, 137, 151
Universal asynchronous receiver/
    transmitters. *See* UART
Update rate, 4

Vector. *See* Interrupt vectors
VME bus, 226–227
von Neumann architecture, 15–16,
    16f

Wait_On, 210

Wait states
    bus types and, 34–35
    description of, 33, 61
    dual-port RAM and, 177
    extended data hold time and, 63
    flash memory and, 238
    integral generators, 34
    internal, 33
    timing of, 33–34, 34f
Watchdog timer
    description of, 75
    electrostatic discharge protection
        secondary to, 82
    functions of, 75
    mechanism of operation, 75, 76f
    sophisticated types of, 75–76
Websites, 307–308
Wide cache memory, 241f

Yield, 210

Z80 peripherals, 60–61
Zilog Z-80, 6, 30f, 31, 225